ALBERT CHARLES SEWARD

CATALOGUE
OF THE
MESOZOIC PLANTS
IN THE
DEPARTMENT OF GEOLOGY
BRITISH MUSEUM
(NATURAL HISTORY)

THE JURASSIC FLORA

II. - LIASSIC AND OOLITIC FLORAS
OF ENGLAND
(EXCLUDING THE INFERIOR OOLITE PLANTS OF THE YORKSHIRE COAST)

PLATES I.-XIII.

Elibron Classics
www.elibron.com

Elibron Classics series.

ISBN 1-4212-4430-6 (paperback)
ISBN 1-4212-4429-2 (hardcover)

This Elibron Classics Replica Edition is an unabridged facsimile of the edition published in 1904 by the British Museum, London.

CATALOGUE

OF THE

MESOZOIC PLANTS

IN THE

DEPARTMENT OF GEOLOGY.

PART IV.

CATALOGUE

OF THE

MESOZOIC PLANTS

IN THE

DEPARTMENT OF GEOLOGY

BRITISH MUSEUM

(NATURAL HISTORY).

THE JURASSIC FLORA.

II.—LIASSIC AND OOLITIC FLORAS OF ENGLAND

(EXCLUDING THE INFERIOR OOLITE PLANTS OF THE YORKSHIRE COAST).

PLATES I–XIII.

BY

A. C. SEWARD, F.R.S.,

UNIVERSITY LECTURER IN BOTANY, FELLOW AND TUTOR OF EMMANUEL COLLEGE, AND
LATE FELLOW OF ST. JOHN'S COLLEGE, CAMBRIDGE.

LONDON:

PRINTED BY ORDER OF THE TRUSTEES.

SOLD BY

LONGMANS AND CO., 39, PATERNOSTER ROW.

B. QUARITCH, 15, PICCADILLY. DULAU AND CO., 37, SOHO SQUARE, W.

KEGAN PAUL AND CO., 43, GERRARD STREET, SOHO, W.

AND AT THE

BRITISH MUSEUM (NATURAL HISTORY), CROMWELL ROAD, S.W.

1904.

HERTFORD

STEPHEN AUSTIN AND SONS,

PRINTERS.

PREFACE.

This second and concluding Part of Mr. Seward's Catalogue of the Jurassic Floras is rendered somewhat unsatisfactory by the very fragmentary nature of nearly all the fossils hitherto discovered. Its value, however, is much enhanced by the care with which the author has compared the more important specimens in other Museums with the collection in the British Museum, which is the immediate subject of his work. The plan adopted makes the Catalogue a nearly complete treatise on the known fossil remains of plants from the Trias, Rhætic, Lias, and Oolite of England; and, with all its necessary imperfections, it forms a secure and valuable basis for future research.

ARTHUR SMITH WOODWARD.

Department of Geology.

March, 1904.

AUTHOR'S PREFACE.

In the volume "The Jurassic Flora. I. The Yorkshire Coast," issued in 1900, I dealt exclusively with specimens in the British Museum and in other collections obtained from the Inferior Oolite plant-beds of East Yorkshire. In the present volume are included plants from Jurassic rocks in various parts of England, together with a few from the Coralline Oolite and Liassic strata of East Yorkshire. A short account is also given of such Triassic and Rhætic plants as are represented in the Museum collection. References will be found in the following pages to Jurassic specimens in several provincial museums.

My work in various museums has been greatly facilitated by the ready help afforded by those in charge of the collections. I wish to offer my thanks to Professor Sollas for his kindness in allowing me access to the rich collection of Stonesfield fossils in the Oxford Museum, and for affording me an opportunity of having several specimens photographed. My thanks are due also to Mr. E. T. Newton for the friendly assistance which he has repeatedly given me in my examination of fossil plants in the Jermyn Street Museum. I am indebted to Dr. Hoyle of Manchester, to Mr. Platnauer of York, to Mr. Newsham of Whitby, and to Mr. Slater of Malton for the loan of specimens. I wish to express my thanks also to Mr. A. M. Bell of Oxford for his courtesy in showing me his private collection of Stonesfield plants, and for lending me the fossil which I have made the type of a new species, *Sphenozamites Belli*.

I must not omit a grateful acknowledgment to the staff of the Geological Department and to Dr. A. B. Rendle and Mr. Gepp of the Botanical Department for their assistance during my visits to the Museum. Mr. Arber's recent work in rearranging and classifying the material in the Fossil Plant Gallery in the Geological Department has done much to lighten the task of recording the Jurassic species represented in the Museum. I am again indebted to Miss Woodward for the care and skill with which she has drawn the illustrations included in this volume.

A. C. SEWARD.

EMMANUEL COLLEGE, CAMBRIDGE.
March, 1904.

NOTE.

The numbers in brackets after the Authors' names in the footnotes refer to the year of publication of the works quoted.

A bibliography at the end of the volume includes the books and papers cited in the text.

LIST OF FIGURES IN THE TEXT.

			PAGE
Fig.	1.	*Carpolithes*, sp.	19
,,	2.	*Thinnfeldia rhomboidalis*, Ett.	34
,,	3.	*Cycadites rectangularis*, Brauns	39
,,	4.	Cast of Coniferous stem	53
,,	5.	Jet and cast of wood	67
,,	6.	Wood and jet	70
,,	7.	Wood and jet	71
,,	8.	*Dictyophyllum*, sp.	90
,,	9.	*Ginkgo digitata* (Brongn.)	99
,,	10.	*Baiera Phillipsi*, Nath.	102
,,	11.	*Zamites megaphyllus* (Phill.)	114
,,	12.	*Sphenozamites Belli*, sp. nov.	120
,,	13–15.	*Carpolithes conicus*, L. & H.	127, 128
,,	16–18.	*Araucarites ooliticus* (Carr.)	135, 136
,,	19.	*Thuites expansus*, Sternb.	145
,,	20.	*Thuites*, sp.	148

CONTENTS.

PAGE

INTRODUCTION 1

I. TRIASSIC SYSTEM 7

GYMNOSPERMÆ.

Carpolithes, sp.

Indeterminable fragments.

II. RHÆTIC SYSTEM 9

EQUISETALES.

Equisetites Muensteri, Sternb.

LYCOPODIALES.

Lycopodites lanceolatus (Brodie).

FILICALES.

Clathropteris platyphylla (Göpp.).

GYMNOSPERMÆ.

Carpolithes, sp.

PLANTA INCERTÆ SEDIS.

? *Araucarites*, sp.

Fragments referred by Buckman to Angiosperms.

III. LIASSIC SYSTEM 22

THALLOPHYTA.

EQUISETALES.

? *Equisetites Muensteri*, Sternb.

FILICALES.

Thinnfeldia rhomboidalis, Ett.

PLANTA INCERTÆ SEDIS.

Ctenopteris cycadea (Brongn.).

CYCADOPHYTA.

Cycadites rectangularis, Brauns.
Otozamites obtusus (L. & H.).
(Cycadean stems.)
Cycadeoidea gracilis (Carr.).
? *C. pygmæa* (L. & H.).

CONIFERALES.

Pagiophyllum peregrinum (L. & H.).

PLANTA INCERTÆ SEDIS.

Strobilites elongata, L. & H.

CONIFEROUS WOOD.

Araucarioxylon Lindleii (With.).
Cupressinoxylon Barberi, sp. nov.

JET.

IV. OOLITIC SERIES 73

THALLOPHYTA.

? *Algites furcatus* (Brongn.).
Girvanella pisolitica, Weth.
Solenopora jurassica, Brown *ex* Nich. MS.
Caulerpa Carruthersi, Murray.

FILICALES.

Laccopteris Woodwardi (Leck.).
Dictyophyllum, sp.
Tæniopteris vittata, Brongn.
Sphenopteris, sp. *a*.
Sphenopteris, sp. *b*.
Sagenopteris Phillipsi (Brongn.).
Cf. *Thinnfeldia speciosa*, Ett.

GINKGOALES.

Ginkgo digitata (Brongn.).
Baiera Phillipsi, Nath.

CYCADOPHYTA.

Williamsonia Bucklandi (Ung.).
W. pecten (Phill.).
Cycadeoidea squamosa (Brongn.
Zamites megaphyllus (Phill.).
Ctenis latifolia (Brongn.).
Otozamites, sp.
Sphenozamites Belli, sp. nov.

A GENUS OF GYMNOSPERMS OF DOUBTFUL AFFINITY.

Podozamites stonesfieldensis, sp. nov.

GYMNOSPERMOUS SEEDS.

Carpolithes conicus, L. & H.
C. diospyriformis, Sternb.
Carpolithes, sp.

CONIFERALES.

Araucarites sphærocarpus, Carr.
A. ooliticus (Carr.).
A. Brodiei, Carr.
A. sphæricus (Carr.).
A. Bucklandi (L. & H.).
A. Cleminshawi, Mans.-Pley.
Araucarites, sp.
Thuites expansus, Sternb.
Thuites, sp.
Brachyphyllum, sp. *α*.
Brachyphyllum, sp. *β*.
Conites primævus (L. & H.).
C. depressus (Carr.).
? *Nageiopsis*, sp.
? *Phyllites*, sp.

CONCLUSION 157

BIBLIOGRAPHY 171

INDEX 185

EXPLANATION OF PLATES.

INTRODUCTION.

The specimens described in a previous volume entitled *The Jurassic Flora*, Part I., published in 1900[1] as one of the series of Catalogues of the Mesozoic Plants in the Department of Geology in the British Museum, were obtained from the rich plant-bearing strata of the Yorkshire Coast. The present volume is devoted to the description of English plants from Rhætic, Liassic, and Oolitic rocks. With the exception of the comparatively large assemblage of forms afforded by the Stonesfield Slate of Oxfordshire, and the smaller number of species obtained from the Liassic rocks of Dorsetshire, the specimens described in the following pages represent scattered relics of various Jurassic floras, and are thus in striking contrast to those which formed the subject of the earlier volume.

The few fragments of plants obtained from British Triassic strata preserved in English Museums are too imperfect to be determined with accuracy, and hopelessly inadequate to form the subject of a separate memoir. In a monograph on the older Mesozoic floras of the United States published in 1900, Professor Lester Ward[2] speaks of the incompleteness of the botanical records of Triassic age in North American strata: our British records are even more fragmentary and unsatisfactory.

There are good reasons for believing that at the close of the Permo-Carboniferous era the plant-world experienced a striking change in its constitution.[3] It is, therefore, particularly unfortunate that our records of Triassic floras are not more abundant.

---

[1] Seward (00). See also Seward (00^2).
[2] Ward (00), p. 221.
[3] Seward (02).

Although an account of Triassic vegetation based on data furnished by extra-British strata would be out of place in the present volume, it is important that brief reference should be made to the nature of the floras immediately preceding those of Rhætic and Jurassic age. The object of a detailed description of the Jurassic plants of Britain is not merely to supply information as to the composition of local floras, but to furnish facts which may be of service in the consideration of questions of wider application.

The general character of the vegetation during the Coal period appears to have persisted into the Permian epoch without any sweeping change. Over a considerable portion of the northern hemisphere the Upper Palæozoic vegetation was characterised by a striking uniformity; an assemblage of plants of similar facies occupied a wide area in China, and traces of the same northern flora have been found as far south as the Zambesi district of Africa.[1] On the other hand, the Permo-Carboniferous forests, of which scattered remnants are preserved in India, South America, South Africa, and Australia, must have presented a very different appearance, due for the most part to the almost complete absence of *Calamites*, *Lepidodendron*, *Sigillaria*, and other common types of the great northern flora. The southern or *Glossopteris* flora is of interest from the point of view of the later northern floras. The occurrence in Lower Triassic rocks of the Vosges of *Neuropteridium* and *Schizoneura*—two characteristic members of the Permo-Carboniferous vegetation in southern regions—affords a connecting link between the Triassic flora in the north and the *Glossopteris* flora of Gondwana Land.[2] It seems clear that a few of the Triassic types migrated to the north at the close of the Palæozoic era, and established themselves in Europe as members of the oldest Triassic vegetation. On the other hand, the Permian floras of Europe contain some genera, such as *Equisetites* and a few Cycadean plants, which may be regarded as pioneer forms foreshadowing the transition from the Palæozoic to the Mesozoic vegetation. Making due allowance for

---

[1] Zeiller (83).

[2] See Seward (97 and 02). See also Zeiller (97) for an excellent account of Permo-Carboniferous floras.

the Imperfection of the Geological Record, the facts unmistakably point to a widespread change in the constitution of northern floras at the close of the Permian era—a change which appears to have been effected with comparative suddenness. As the result of certain circumstances, evolutionary forces acted with increased rapidity at this stage in the history of the plant-world. There is always a danger in speaking of the change from one flora to another as having been abrupt or sudden, as the expression implies an occurrence at variance with our ideas of the course of action of evolutionary forces. It must be remembered that the documents from which the history of plant-life is compiled are exceedingly fragmentary, and we cannot form any conception as to the interval which separated one flora from another. In stating that the character of the vegetation appears to have undergone an almost sudden change, we simply mean that within a short period, as measured by geological records, sweeping changes were effected in the composition of the vegetation.

There can be no reasonable doubt that at certain stages in the evolution of floras there was a widespread alteration in the relative preponderance of classes. It would seem that the march of plant-life has been characterised by long-continued periods, during which no fundamental changes were effected as regards the balance of power among the different classes of plants, succeeded at rare intervals by stages which were distinguished by striking and comparatively rapid reorganisation, and by an alteration in the dominant type. The older Triassic floras introduce us to a new type of vegetation, which, with gradual additions and modifications, persisted up to the end of the Wealden period. In the southern hemisphere the Permo-Carboniferous floras differ but slightly from those of the Triassic epoch; in India, for example, the plant-beds of the Damuda and Panchet series have afforded evidence of comparative continuity between the later Palæozoic and the earlier Mesozoic vegetation. As we pass up to the succeeding Rajmahal series a vegetation is met with which in essential features agrees with that of Rhætic and Liassic age in the northern hemisphere. It would seem, then, that in the south the Mesozoic type of vegetation appeared at a somewhat later period than in the north. There must have been a considerable interval between the deposition of the Upper Damuda beds of

India and the lower members of the Rajmahal series;[1] this is inferred not only from the well-defined differences between the two floras, but from the greater amount of disturbance to which the Damuda rocks have been exposed. The marked resemblance between the plants of the Rajmahal series of India and those from the Stormberg beds of South Africa, with the Upper Triassic and Rhætic floras of the north, demonstrates an extraordinary uniformity in the older Mesozoic vegetation, which extended from Northern Europe and America to Africa, India, Australia, and China. This uniformity appears to have persisted throughout the greater part of the Mesozoic era until, with the advent of Angiosperms, drastic and widespread changes were once more effected in the floras of the world.

It is much to be regretted that our knowledge of Lower Triassic vegetation in Europe is extremely small; owing to the paucity of evidence furnished by the strata belonging to this critical period in the development of the plant-world, we are unable to follow the steps by which the Mesozoic facies was established. The collection of plants from the Bunter beds of the Vosges, originally described by Schimper & Mougeot in 1844,[2] supplies the greater part of our knowledge of the earlier Triassic vegetation in Europe. We must not, however, lose sight of such connecting links as exist between the Permo-Carboniferous and Triassic floras. The genus *Neuropteridium*, a plant which may be a true fern, or possibly a surviving member of the Cycadofilices, is represented in the Triassic plant-beds of the Vosges by a species which can hardly be distinguished from one which flourished in South America, South Africa, and India in the Palæozoic era. Similarly, the genus *Schizoneura* is represented in the European Triassic vegetation by a type which cannot be distinguished by any important features from *S. gondwanensis* from the Lower Gondwana rocks of India. It would seem, then, that some members of the southern Permo-Carboniferous vegetation had become established in Europe during the earlier part of the Triassic era. The genus *Pleuromeia*, which makes its appearance in Triassic rocks, is known

---

[1] Medlicott & Blanford (new edition by R. D. Oldham), (93), p. 177.
[2] Schimper & Mougeot (44).

only in the form of casts exhibiting a strong likeness to some Palæozoic Lycopods. This plant has been found in Bunter rocks of Germany,[1] and Fliche[2] has recently recorded its occurrence in strata referred to the upper portion of the Middle Trias at Chauffontaine, near Lunéville. Professor Fliche, in a note published in the *Comptes Rendus*,[2] also records the occurrence in Bunter beds near Baccarat of a type which he names *Stigmarites Nicklesi*, and describes as probably generically identical with *Stigmaria ficoides*. The same author is of opinion that the fossil named by Schimper & Mougeot *Caulopteris tessellata* should be placed among the Lycopodinæ, and referred either to *Lepidodendron* or to a new genus closely allied to that characteristic member of Palæozoic floras. The somewhat doubtful *Sigillaria oculina*,[3] first recorded by Blanckenhorn[4] from the Trias of Commern, in the Rhine Province, is in itself insufficient as proof of the existence of the true *Sigillaria* in the Triassic era. From Middle Triassic strata a few plants have been described from Recoaro, in Italy,[5] and from a few other localities; but the Muschelkalk, like the underlying Bunter series, represents a chapter in the history of plant-evolution of which but a few fragmentary pages have been discovered. The greater part of the Triassic sediments in Europe are of marine origin, and from them have been obtained several species of calcareous algæ referred to the ancient family Siphoneæ.[6] On the other hand, the Triassic beds of Britain and other parts of Europe, as shown by the abundance of saline deposits and by certain lithological characters, were formed at a time when the conditions were not such as to favour the luxuriant development of plant-life.

Passing to the higher members of the Triassic system, we are suddenly in touch with a rich Mesozoic vegetation, which continues with little modification through the whole of the Jurassic era. The differences between the Keuper floras and those from the overlying Rhætic strata are very slight, and for general

---

[1] Solms-Laubach (99), p. 241; Potonié (99), p. 217.
[2] Fliche (03).
[3] Potonié (99), p. 256, fig. 246.
[4] Blanckenhorn (86); Weiss (86).
[5] Benecke (68).
[6] De Lorenzo (97); Benecke (98).

purposes we may group together the floras of Virginia,[1] Taylorsville (California),[2] the Lettenkohle and Lunz floras of Austria, the rich floras of Scania,[3] Franconia,[4] and other European regions,[5] as so many scattered samples of a luxuriant vegetation characterised by its worldwide distribution and uniformity of composition.

---

[1] Fontaine (83).
[2] Ward (00).
[3] Nathorst (78–86), (78[2]), (78[3]).
[4] Schenk (67).
[5] Heer (76), Compter (74), Jaeger (27), Stur (85), Schenk (66), etc.

I.—TRIASSIC SYSTEM.

The plant fragments of Keuper age preserved in the British Museum collections do not strictly fall within the scope of the present volume, but as they are too small and obscure to form the subject of a separate memoir they may be briefly enumerated. In the Jermyn Street Museum there are also a few indistinct specimens of Keuper plants from Bromsgrove and Pendock in Worcestershire and from Rowington in Warwickshire,[1] but their preservation is not sufficiently good to admit of accurate determination.

Class GYMNOSPERMÆ.

Genus **CARPOLITHES**, Sternberg.

[Flor. Vorwelt, iv. p. xl. 1823.]

Carpolithes, sp.

V. 1077. A very obscure impression of a seed: indeterminable. Longdon (Keuper), Worcestershire.

Presented by Rev. P. B. Brodie, 1885.

V. 2530. A similar specimen from Rowington, Warwickshire.

Presented by Rev. P. B. Brodie, 1891.

V. 3389, V. 3395, V. 3398. Seeds from Rowington and Pendock. An example from the Keuper of Rowington (V. 3398) is 2 cm. in length, and has the form of a cordate convex cast surrounded by a flat border. *Brodie Coll.*

V. 3640. Small seeds. *Slatter Coll.*

---

[1] Brodie (87).

INDETERMINABLE FRAGMENTS.

The following specimens, some of which are no doubt fragments of Coniferous plants (*Voltzia*?), are too imperfect to describe or name:—

V. 3386, V. 3387, V. 3388. Plant fragments from Rowington.
Brodie Coll.

V. 3389. A specimen labelled "*Voltzia* in fruit." Too indistinct to determine.
Rowington. *Brodie Coll.*

V. 3390. Possibly a portion of a cone.
Rowington. *Brodie Coll.*

V. 3392, V. 3393. Fragments from Longdon, near Pendock, Worcestershire, and from Rowington, Warwickshire. *Brodie Coll.*

24,190. Half-relief tracks or castings, labelled "*Gorgonia Keuperi*," from the Keuper Sandstone of Leicester.
Presented by James Plant, Esq., 1849.

II.—RHÆTIC SYSTEM.

The sediments between the uppermost Keuper beds and the lowest strata of the Lias are by many authors included in the Triassic system,[1] but by others the Rhætic series is regarded as forming the base of the Jurassic system.[2] The occurrence of particularly rich and well-preserved floras in beds of Rhætic age near Baireuth, Bamberg, and other localities in Franconia, in the coal-bearing strata of Scania, and in other regions, has brought this period into prominence as one that is of special importance from a palæobotanical standpoint. In Britain the Rhætic strata form a thin band stretching diagonally across England from near Redcar on the Yorkshire coast to the cliffs of Dorsetshire. Rocks of this period are probably represented in the outlying patch of Lower Lias beds to the west of Carlisle;[3] Rhætic beds occur also in the north-east of Ireland and in the north of Scotland.

The vegetation of the land bordering the shallow Rhætic sea is represented in the British area by a few fragmentary fossils, which afford a striking contrast to the richness of the plant-beds of this period in other parts of the world.

The line of division between the base of the Lias and the summit of the Rhætic is not very well defined, and the difficulty of distinguishing between beds of Upper Rhætic and of Lower Liassic age is increased by the common use of a misleading terminology. The Rhætic series is often subdivided into three stages, the White Lias at the top, succeeded by the Black shales and Grey marls. The term White Lias, as pointed out by Mr. E. Wilson,[4] is an unfortunate one; it has been applied

---

[1] Geikie (93), p. 867; Blanford (84), p. 318.
[2] De Lapparent (00).
[3] Holmes (81), (99).
[4] Wilson, E. (91).

by different authors to different things, both Lias and Rhætic beds having been included in this designation. The term Upper Rhætic has been suggested as preferable to that of White Lias for the uppermost beds of the series.

In Murchison's *Geology of Cheltenham*,[1] an Insect Limestone of Lias age is described as having afforded many remains of insects, ferns, and the leaves "of a new plant belonging to the family Naiadeæ or some of its allies." The species *Naiadita acuminata*, founded by Brodie,[2] is recorded from an insect limestone of 'Lower Lias' age at Wainlode Cliff, Tewkesbury. The term Insect Limestone was applied by Brodie to bands of limestone at the base or near the base of the Lower Lias or at the summit of the Rhætic series.

In 1850 Mr. J. Buckman[3] communicated a paper to the Geological Society *On some Fossil Plants from the Lower Lias*; his object being to describe the vegetable remains associated with the insects in the 'Lower Lias' beds, and "to distinguish the new forms which have been made out since the publication of the *History of Fossil Insects*." Buckman gives the following list of plants from strata which he classes as Lower Lias, but which should rather be referred to the Upper Rhætic series.

Equisetum Brodiei, Buck., fig. 1.[4] (Type in the British Museum, No. V. 3358.) = *Equisetites Muensteri*, Sternb.
Otopteris obtusa, L. & H. = *Otozamites obtusus* (L. & H.).
O. acuminata, L. & H. = *Otozamites acuminatus* (L. & H.).
Naiadita lanceolata, Brodie, fig. 2 } = *Lycopodites lanceolatus* (Brodie).
N. petiolata, Buck., fig. 4 }
Cupressus (?) *latifolia*, Buck., fig. 5 = ? *Pagiophyllum peregrinum* (L. & H.).
Hippurites (?), figs. 6 and 7. (V. 3401, Brit. Mus.) } Indeterminable fragments.
Umbelliferæ (?), Traces of, fig. 8. (V. 3585, Brit. Mus.) }
Ericaceæ (?), Traces of, fig. 9. (V. 3586, Brit. Mus.) }

The species *Otozamites obtusus* (L. & H.) is recorded by Brodie from the Insect Limestone of Ashelworth and from Brockeridge; *Otozamites acuminatus* (L. & H.) is also doubtfully recorded by the same author. I have not seen specimens of either species

[1] Murchison (45), p. 52.
[2] Brodie (45), p. 93.
[3] Buckman (50).
[4] The figures referred to are given in Buckman's paper of 1850.

from an undoubted Rhætic locality. The specimens referred by Buckman to the Angiosperms, and doubtfully spoken of as *Hippurites* and as traces of Umbelliferæ and Ericaceæ, are of no botanical value and cannot be determined. In a paper by Sorby published in vol. vii. of the Transactions of the Microscopical Society of London, a specimen of wood is described from the Lias of Keynsham, near Bristol.[1] Sorby's figure shows that the structure is that of a typical Dicotyledon; the fossil was purchased from a dealer, and the locality is given with some hesitation. It is in the highest degree improbable that the wood was obtained from Rhætic rocks.

The list of Rhætic fossils given by Mr. H. B. Woodward[2] in the Geological Survey Memoir on the East Somerset and Bristol Coal-fields includes three plants—*Fucoides, Naiadita acuminata* (?), Buck., and *Equisetites*. The name *Naiadita acuminata* was used by Buckman in his description of new species of plants included in Murchison's *Geology of Cheltenham*;[3] it was no doubt applied to the same plant which Buckman also named *Naiadita lanceolata*.

In 1891 Mr. E. Wilson[4] published a detailed description of an unusually good section of Rhætic beds exposed in a deep railway cutting at Pylle Hill, Totterdown, Bristol. He mentions the occurrence of *Naiadita* in a bed of limestone in the Upper Rhætic series, and points out that the plant was spoken of by Brodie and Buckman as having been derived from the Lias. A redescription of the Pylle Hill section was published in 1900 by Mr. Wickes, who speaks of one of the Rhætic beds as "being in some zones full of the little water-plant *Naiadites acuminatus* in a fine state of preservation."[5] The plant-genus *Naiadita*, which has long been regarded as of uncertain botanical position, has recently been carefully examined by Miss Igerna Sollas,[6] who brings forward good evidence in support of her conclusion that *Naiadita lanceolata* must be referred to the Lycopodiaceæ.

---

1 Sorby (52).
2 Woodward, H. B. (76), p. 90.
3 Murchison (45), p. 93.
4 Wilson (91).
5 Wickes (00), p. 422.
6 Sollas, I. B. J. (01).

Class EQUISETALES.

Genus **EQUISETITES**, Sternberg.

[Flor. Vorwelt, v.–vi. p. 43, 1833.]

Equisetites Muensteri, Sternberg.

(Pl. I. Fig. 4.)

1838. *Equisetites Muensteri*, Sternberg, Flor. Vorwelt, pl. xvi. figs. 1–5.
1850. *Equisetum Brodiei*, Buckman, Quart. Journ. Geol. Soc. vol. vi. p. 414, fig. 1.
1854. *E. Brodiei*, Morris, Brit. Foss. p. 8.
1867. *Equisetites Muensteri*, Schenk, Foss. Flor. Grenz. p. 14, pl. ii. figs. 3–9; pl. iii.
1873. *E. Muensteri*, Saporta, Pal. Franç. vol. i. p. 232, pls. xxvii.–xxix.
1891. *E. Muensteri*, Krasser, Sitzber. k. Akad. Wiss. Wien, vol. c. Abth. 1, p. 10.
1892. *E. Muensteri*, Bartholin, Bot. Tid. Bot. For. Kjöbenhavn, vol. xviii. Heft 1, p. 13, pl. v. figs. 1–5.
1893. *E. Brodiei*, Woodward, Jurassic Rocks, iii. p. 378.
1894. Cf. *Equisetum pseudo-hoerense*, Saporta, Flor. Portugal, pl. i. figs. 1–4.
Cf. *Equisetum Renaulti*, Raciborski, Flora Kopalna, p. 231, pl. xxvii. figs. 1–14.
1896. *Equisetum Muensteri*, Hartz, Meddelelser om Grönland, xix. p. 223, pl. vi.
1902. *E. Muensteri*, Möller, Bornholms Flor. p. 58, pl. vi. figs. 13–15.

Aerial shoots characterised by the comparatively small number of internodal ridges and by the few and short teeth of the leaf-sheaths. Strobilus, as figured by Schenk,[1] of the normal Equisetaceous type.

There can be no doubt that this single specimen of an Equisetaceous plant from Worcestershire should be referred to *Equisetites Muensteri*, a species characteristic of Rhætic floras in various parts of the world. Schenk's figures of this type from the Rhætic of Franconia clearly demonstrate—so far, at least, as external vegetative features are concerned—the identity of the English and German forms. Similarly, as shown by

[1] Schenk (67), pl. iii.

Bartholin, Möller, and Hartz, the same species occurs in Bornholm and East Greenland. The plant described by Raciborski as *Equisetum Renaulti* from Poland belongs either to Sternberg's species or represents a closely allied form. Specimens from Rhætic rocks of Tonkin recently figured by Zeiller [1] and referred by him to a distinct species, *Equisetum Sarrani*, point to the occurrence in the far East of a closely allied, if not identical type.

V. 3358. Pl. I. Fig. 4.

Type-specimen of Buckman's *Equisetum Brodiei* (Quart. Journ. Geol. Soc. vol. vi. p. 414; figure $\frac{1}{7}$ natural size). The specimen is 20 cm. long and 1 cm. broad, with three nodes, two of which are shown in the portion represented in Fig. 4. The broad internodal ribs and the few broad leaf-teeth suggest the reference of this specimen to the species *Equisetites Muensteri*, [2] an identification first made by Carruthers, to whom the fossil was submitted in 1889. Buckman describes the specimen as fertile, but there is no trace of any strobilus.

Strensham, Worcestershire. *Brodie Coll.*

V. 774. Two carbonaceous bodies, which may be nodal diaphragms of an Equisetaceous stem: too obscure to determine.

Purchased, 1885.

Class LYCOPODIALES.

Genus **LYCOPODITES**, Brongniart. [3]

[Prodrome, p. 83, 1828.]

This generic name is used in a comprehensive sense as including fossils which may belong either to the heterosporous Selaginellaceæ or to the homosporous Lycopodiaceæ. In view of the careful examination by Miss Sollas of specimens of the plant formerly described as *Naiadita lanceolata*, I have adopted the name *Lycopodites* as the more fitting designation.

---

[1] Zeiller (02), pl. xxxix.

[2] Cf. Schenk (67), pls. ii. and iii. Compare also *Equisetum Sarrani*, Zeill., Zeiller (02), pl. xxxix.

[3] See Seward (00), p. 68.

Lycopodites lanceolatus (Brodie).

(Pl. II. Figs. 2 and 3.)

1845. *Naiadita lanceolata*, Brodie, Fossil Insects, p. 93.
Naiadea acuminata, Buckman, in Murchison's Geol. Cheltenham, p. 6.
1850. *N. lanceolata*, Buckman, Quart. Journ. Geol. Soc. vol. vi. p. 415, fig. 2.
1888. *N. lanceolata*, Prestwich's Geology, p. 168, pl. lxxvi *a*.
1900. *Naiadites acuminatus*, Wickes, Proc. Geol. Assoc. vol. xvi. pt. 7, p. 422.

Plant slender and moss-like in habit. The axis, which is delicate and thread-like, bears numerous linear acuminate or narrow ovate leaves reaching a length of approximately 5 mm. Under a low magnifying power the thin lamina of the leaves is seen to be made up of a layer of polygonal or rectangular cells arranged in parallel vertical series; there is no trace of a midrib and no stomata. The sporangia are more or less spherical and short-stalked, the larger ones being about 0·75 mm. in diameter; they are situated at the bases of the leaves, and contain numerous tetrads of spores. The spores measure 0·08 mm. in diameter.

In a *Notice on the discovery of the remains of Insects in the Lias of Gloucestershire*, published in 1842, Mr. Brodie mentions the occurrence of numerous small plants in certain beds at Wainlode Cliff, on the Severn, some of which he compares with mosses. The same author, in his work on *Fossil Insects*, proposed the name *Naiadita lanceolata* for one of the plants from the Insect Limestone, the generic name being chosen as the result of Professor Lindley's determination of the leaves as being those of a Monocotyledon.

Brodie mentions that the best specimens of *Naiadita* were found in abundance on the Wells road, near Bristol. No complete diagnosis accompanies the reference to the species. In Murchison's *Geology of Cheltenham* the genus *Naiadea* is thus defined:—"Parallel veined endogenous plants, having the appearance of aquatics; leaves varying somewhat in form on the same plant, and presenting in their character a close affinity to the recent family Naiadaceæ."[1]

[1] Murchison (45), p. 93.

The species *Naiadea acuminata*, which is no doubt the plant named by Brodie *N. lanceolata*, is defined as follows:—"Upper leaves ovate lanceolate, lower ones much acuminated. A small plant, having a great resemblance to our recent species of *Callitriche*."

Buckman's list of plants in the appendix to Murchison's volume includes two other species under the generic name *Naiadea*, *N. obtusa* and *N. ovata*;[1] the species so named are from the Stonesfield Slate, and have no family or generic affinity with the species from Wainlode Cliff. The Stonesfield fossils are referred to in the sequel as leaflets of a Gymnospermous plant, probably *Podozamites*. We are here concerned only with the species *Naiadea lanceolata*. In 1886 Mr. Starkie Gardner[2] examined some specimens in Mr. Brodie's collection which are now in the British Museum, and suggested that they represented fragments of a moss comparable with the recent fresh-water *Fontinalis*. In this opinion he was supported by Mr. Carruthers and Mr. Murray of the British Museum. In a footnote to his paper Gardner mentions a moss-capsule received from Mr. Brodie from the same beds which had afforded the specimens of the slender stems and leaves. As I have already stated,[3] the supposed moss-capsule is too obscure to determine. The specimen (V. 3584) has the form of an oval brown stain on the surface of the rock, with a suggestion of a stalk at one end, but there are no adequate grounds for regarding it as a capsule.

In a paper communicated to the Geological Society in 1901, Miss Igerna Sollas gave an account of a detailed investigation of several specimens of *Naiadita* from different localities in the Severn Valley. The best material obtained in recent years has been collected by Mr. W. H. Wickes of Bristol, to whom I am indebted for the loan of numerous specimens. The plant occurs in a fragmentary state, and monopolises the surface of fairly large slabs of rock. Miss Sollas accurately describes it as "delicate, slender, and moss-like in habit . . . . some of the strata contain loose leaves and disconnected pieces of stem only, while in

---

[1] Murchison (45), pp. 93, 94, pl. i. fig. 2; pl. ii. fig. 1.

[2] Gardner (86), p. 203. See also Seward (98), p. 240.

[3] Seward, loc. cit.

others the stems may branch." Sporangia are fairly common, and occasionally occur in organic connection with the leaf-bearing stem.

Miss Sollas submitted several sporangia to a minute examination, and found that it was possible to dissolve out the spores by dilute hydrochloric acid. The spores, which are tetrahedral and provided with two coats, a smooth intine and an exine covered with irregular bosses, measure 0·08 mm. in diameter, being twice as large as the spores of any recent species of *Lycopodium* examined by Miss Sollas. The stem is covered with an epidermis composed of long rectangular cells, and below this were found several narrow tubular elements. In the leaves only a single layer of cells was detected, but Miss Sollas considers that the lamina was originally made up of several layers, while stomata, of which no trace was seen in the epidermal layer of the leaf, may have been present. Without quoting further from Miss Sollas' paper, which should be consulted for a more complete account, I may add that I entirely share the conclusion arrived at, that *Naiadita* must be regarded as a member of the Lycopodiaceæ. This Rhætic species differs in its much more slender habit and in the narrow form and smaller size of its leaves from the Inferior Oolite type *Lycopodites falcatus*, L. & H.[1]

V. 4015. Pl. II. Figs. 2 and 3.

The fragment represented in Fig. 2 illustrates the slender moss-like habit of the plant; Fig. 3 shows the single layer of epidermal cells. Under this number (V. 4015) are included several slabs of rock covered with broken fragments of stems and leaves of *Lycopodites lanceolatus*.

Near Redland. *Presented by W. H. Wickes, Esq.*

V. 3356. Figured by Buckman as *Naiadita petiolata*, Quart. Journ. Geol. Soc. p. 415, fig. 4, 1850. The figure is not very accurate, and represents the plant much larger than natural size.

Estheria bed, near Bristol. *Brodie Coll.*

V. 3357. [Said to be figured by Buckman, Quart. Journ. Geol. Soc. figs. 2–3, 1850.] Fragments like that shown in Pl. II. Fig. 2.

Near Bristol. *Brodie Coll.*

Other specimens :—**V. 1437, V. 3394, V. 3397.**

---

[1] Lindley & Hutton (31), pl. lxi.

Class FILICALES.

Family DIPTERIDINÆ.

In my former volume on Jurassic plants this family-name was used to include such fossil genera as *Dictyophyllum*, *Protorhipis*, *Clathropteris*, and others, as well as the four existing species of the genus *Dipteris*, an Indian and Malayan type, which have been placed in a separate family—Dipteridinæ—on account of certain peculiarities which distinguish it from the true Polypodiaceæ.[1]

Genus **CLATHROPTERIS**, Brongniart.

[Prodrome, p. 62, 1828.]

This genus was instituted by Brongniart in 1828 for a plant which he had previously[2] described from the Rhætic rocks of Scania as *Filicites meniscoides*. The most striking characteristics are the pinnatifid fronds and the regular rectangular meshes formed by the finer veins in the lamina. The best known species of the genus is *Clathropteris platyphylla* (Göpp.), of Rhætic age, the same type which Brongniart described in 1825 as *Filicites meniscoides*. *Clathropteris* fronds bear a close resemblance to those of *Dictyophyllum*, a genus which is abundantly represented in the Inferior Oolite flora of Yorkshire, and is a prominent member of the Rhætic floras of Scania and other regions. As I have elsewhere pointed out, the main distinction between the two genera consists in the more regular and rectangular form of the meshes, formed by the secondary and tertiary veins, in the lamina of *Clathropteris*. In all probability the two names have been applied to ferns that are generically identical, and it would probably be more in accordance with the facts of relationship if we absorbed *Clathropteris* into *Dictyophyllum*. It has, however, long been customary to make use of both generic names, and as a matter of convenience we may in this instance retain the generic designation *Clathropteris*.

---

[1] For an account of the Dipteridinæ, see Seward & Dale (01); see also Seward (03[2]).

[2] Brongniart (25), p. 207, pl. xi.

Clathropteris platyphylla (Göppert).

[Gattungen foss. Pflanz. p. 154, pls. xviii.-xix. 1841.]

(Pl. III. Fig. 2.)

1828. *Clathropteris meniscoides*, Brongniart, Hist. vég. foss. pl. cxxxiv. fig. 3.
1832. *Juglandites castaneæfolius*, Berger, Verstein. Sandst. Coburger Gegend. p. 20, pl. iv. fig. 2.
1838. *Camptopteris Muensteriana*, Sternberg, Flor. Vorwelt, p. 168, pl. xxxiii. fig. 9.
1841. *C. platyphylla*, Göppert, Gatt. foss. Pflanz. p. 154, pls. xviii.-xix.
1849. *Clathropteris platyphylla*, Brongniart, Tableau, p. 32.
1867. *C. platyphylla*, Schenk, Foss. Flor. Grenz. p. 81, pls. xvi.-xvii.
1869. *C. platyphylla*, Schimper, Trait. pal. vég. vol. i. p. 636, pl. xlii. figs. 1-3.
1873. *C. platyphylla*, Saporta, Plant. Jurass. Paléont. Franç. [ii.], vol. i. p. 333, pls. xxxvi.-xl.
1876. *C. reticulata*, Heer, Foss. Flor. Helvet. p. 73, pl. xxv. figs. 4-6.
1878. *C. platyphylla*, Nathorst, Floran vid Höganäs, pp. 15, 48, pl. ii.
C. platyphylla, Nathorst, Floran vid Bjuf, p. 41, pl. v. fig. 6; pl. vii. fig. 2.
1882. *C. platyphylla*, Zeiller, Examen Flor. foss. Tong-King, p. 16, pl. x. figs. 12-13; pl. xii. fig. 5.
1892. *C. platyphylla*, Bartholin, Bot. Tid. vol. xviii. p. 26, pl. xi. figs. 1-3.
1893. *C. platyphylla*, Woodward, Lias, p. 378.
1900. *C. platyphylla*, Zeiller, Eléments Paléobot. p. 116, figs. 89-90.
1901. *Dictyophyllum platyphyllum*, Seward & Dale, Phil. Trans. R. Soc. vol. cxciv. p. 505.
1902. *Clathropteris platyphylla*, Möller, Bornholms Foss. Flor. p. 46, pl. iv. fig. 14.
C. platyphylla, Zeiller, Flor. foss. Tonkin (Atlas), pls. xxvii.-xxxiv. pl. lvi.

Frond palmate, petiolate; lamina deeply pinnatifid, consisting of several palmately disposed broad linear segments, which are concrescent at the base of the lamina. The segments, which have a serrate edge, are traversed by a stout median rib, from which numerous parallel veins are given off at a wide angle; the lateral veins are connected by tertiary veins, so disposed as to divide the surface of the lamina into more or less rectangular or polygonal areas, which are occupied by a reticulum of more slender veins.

In habit and venation the fronds resemble recent species of the Malayan and Indian genus *Dipteris*.

Sori numerous, circular, composed of a small number of annulate sporangia.

The best figures of *Clathropteris* are those recently published in Zeiller's excellent Atlas of Tonkin plants;[1] drawings of the sori and sporangia are given in Schenk's work on the Rhætic plants of Franconia.

V. 3399. Pl. III. Fig. 2.

A fragment exhibiting the characteristic venation characters of Göppert's widely distributed Rhætic species. The torn piece of lamina is traversed by several secondary veins which spring from one of the main ribs of the frond, and these are connected by tertiary veins.

Bristol. *Brodie Coll.*

GYMNOSPERMÆ.

Genus CARPOLITHES.

Carpolithes, sp.

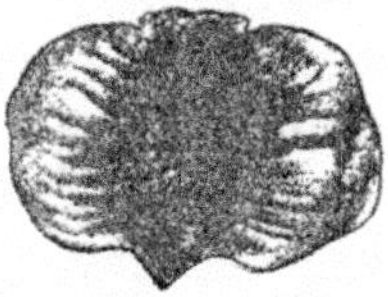

FIG. 1.—*Carpolithes*, sp. V. 3400. Nat. size.

V. 3400. Text-fig. 1.

This specimen, represented natural size in the figure, may be the impression of a winged seed. It consists of a darker central region, with two broad and thin lateral expansions of lighter colour, which are probably membranous wings. The wings exhibit transverse ridges, which may be due to the contraction and

---

[1] Zeiller (02), pls. xxvii.-xxxiv.

wrinkling of their substance. A comparison may be suggested with the broadly winged seeds of the recent *Welwitschia*,[1] an isolated genus of tropical Africa, and the only representative of the Welwitschoideæ, a subdivision of the Gnetales. The specimen is, however, too obscure to determine, and can only be spoken of as possibly a winged Gymnospermous seed referable either to the Coniferales or to the Gnetales. A seed of similar form has been figured by Nathorst from the Rhætic of Scania as *Samaropsis Zignoana*, Nath.[2]

Brodie Coll.

PLANTA INCERTÆ SEDIS.

? **Araucarites**, sp.

V. 3359. A single specimen, which may possibly be an imperfectly preserved cone. It is placed with considerable hesitation in the genus *Araucarites*.

Insect-bed, Binton, Warwickshire. *Brodie Coll.*

INDETERMINABLE FRAGMENTS.

[Referred by Buckman to the Angiosperms.]

The list of plants "from the Lower Lias" given by Mr. Buckman in his paper of 1850[3] includes three specimens referred with some doubt to the Umbelliferæ, Ericaceæ, and to the genus *Hippurites*. It is important to examine with care any possible indications of Angiospermous plants in pre-Cretaceous rocks, and to draw attention to recorded instances of supposed Flowering plants that are botanically valueless. There need be no hesitation in dismissing Buckman's fossils as affording no evidence justifying their reference to the Flowering plants. The following specimens in the Museum collection are too obscure and fragmentary to determine, and, while asserting that they are not the remains of Dicotyledons or Monocotyledons, I am unable to express an opinion as to their nature.

---

[1] See figures in Hooker (63).
[2] Nathorst (78–86), pl. xxv. figs. 10–14.
[3] Buckman (50).

V. 3401. Quart. Journ. Geol. Soc. vol. vi. 1850, p. 415, fig. 7. Figured by Buckman as "*Hippurites*?" The published drawing represents the specimen natural size (not twice natural size, as stated in Buckman's paper). The fragment consists of a stout and apparently woody axis covered with long delicate hairs; it bears a fairly close resemblance to certain scales or hairy bracts of Cycads that have been described in a fossil condition as *Cycadolepis*, but it may be a portion of a root or rhizome.

Brodie Coll.

V. 3585. Quart. Journ. Geol. Soc. vol. vi. p. 416, fig. 8. Described by Buckman as a representative of the Umbelliferæ (?). The specimen, as shown in Buckman's figure, represents a slender branched axis, but it is too small and imperfect to determine.

Brodie Coll.

V. 3586. Quart. Journ. Geol. Soc. vol. vi. p. 417, fig. 9. This specimen has the form of a small leaf 8 mm. in length; the lateral veins represented in Buckman's figure cannot be recognised. The fragment, which was provisionally referred to the Ericaceæ, is absolutely indeterminable.

Brodie Coll.

III.—LIASSIC SYSTEM.

The shallow water in which the Rhætic sediments were deposited became gradually converted into a deeper sea, on the floor of which were accumulated the regular Liassic strata of limestones and shales. Liassic rocks stretch across England from the mouth of the Tees to the coast of Dorset, and occur as isolated patches in Shropshire and Cumberland; exposures of Liassic beds are met with also in Sutherlandshire, in the west of Scotland, and on the Antrim coast. The strata belonging to this system which form the bold headland of Huntcliff, near Saltburn, and the cliffs of Runswick Bay and of other parts of the Yorkshire coast, are essentially marine in origin. Blocks of wood converted into jet in the 'jet-rock' of the Upper Lias, and fragments of silicified wood from a lower horizon, serve as records of the vegetation that flourished on the land bordering the Liassic sea. The Lower Lias strata of Dorsetshire exposed in the cliffs near Lyme Regis are rich in remains of marine life, and from them have also been obtained the majority of British Liassic plants. It is a common occurrence to find an abundance of land-plants entombed in marine sediments, and even in strata that have been deposited in comparatively deep water we meet with traces of terrestrial forms. In a short account of dredging operations in the Eastern Pacific contributed to *Nature* in 1892,[1] Alexander Agassiz speaks of an immense amount of vegetable matter having been dredged from depths of 1,500 fathoms; there was hardly a haul without much water-logged wood and more or less fresh twigs, seeds, fruits, etc., in all stages of decomposition. It is under similar conditions to these that most of the British Liassic plants have been preserved. The Lias strata are usually

[1] Agassiz (92).

conformable to the Rhætic,[1] and thus it is not always a simple matter to decide between Liassic and Rhætic as the most appropriate designation for plants obtained from strata between the uppermost Keuper and the lowest members of the true Jurassic. For present purposes I have included under the term Rhætic the specimens obtained from the neighbourhood of Bristol, confining the designation Liassic to the Dorsetshire and Yorkshire plants. From a botanical point of view, there appear to be no very obvious differences, as regards general composition, between Rhætic and Liassic floras.

In 1824[2] De la Beche published some *Remarks on the Geology of the South Coast of England, from Bridport Harbour, Dorset, to Babbacombe Bay, Devon*, accompanied by a coast-section of the cliffs from Bridport to Sidmouth. He speaks of blocks of fossil wood from Lias limestones, and mentions the rare occurrence of 'ferns' in the rocks near Lyme; two specimens are figured from Buckland's collection, both of which are portions of *Otozamites* fronds collected at Axminster in Dorsetshire. In a later volume the same author published a fuller account of the Dorsetshire coast-section; in this memoir he figured a piece of knotted drift-wood and a fragment of a plant which is, no doubt, *Pagiophyllum peregrinum*.

The following species of Lias plants are described in the *Fossil Flora* of Lindley & Hutton:—[3]

Araucaria peregrina, L. & H. Pl. lxxxviii. 1833. Lyme Regis.	= *Pagiophyllum peregrinum* (L. & H.).
Strobilites elongata, L. & H. Pl. lxxxix. 1833. Lyme.	Indeterminable.
Otopteris obtusa, L. & H. Pl. cxxviii. figs. 1–2. 1834. Membury, near Axminster, and Polden Hill, near Bridgewater, Somersetshire. The type-specimen (fig. 1) is in the Oxford Museum.	= *Otozamites obtusus* (L. & H.).
Cycadeoidea pygmæa, L. & H. Pl. cxliii. 1835. Lyme.	= *Cycadeoidea pygmæa*, L. & H.

[1] Woodward, H. B. (93), p. 22.
[2] De la Beche (24).
[3] Lindley & Hutton (31–37).

A detailed account of the Lias of Yorkshire was published in 1876 by Tate & Blake;[1] they mention the following plants:—

Driftwood from various localities and horizons.
Pecopteris, sp., from the *Ammonites angulatus* zone.
Equisetites (?) *liassicum* from the *A. Bucklandi* zone.
Pachyphyllum peregrinum from the zones of *A. annulatus* and *A. serpentinus*.
Peuce Huttoniana } from the Upper Lias, Whitby.
P. Lindleyana }
Chordophyllites cicatricosus } zone of *A. spinatus*.
Nulliporites furcillatus }

These supposed Algæ are too obscure to be determined, and cannot be accepted as trustworthy records of Liassic plants.

The account given by Tate & Blake of the jet from the Upper Lias beds is referred to in the sequel. The two species of Conifers, *Peuce Huttoniana* and *P. Lindleyana*, were described and figured by Witham in his well-known book of 1833.[2] Much information as to the Liassic rocks of Yorkshire is given by Fox-Strangways[3] in a Survey memoir published in 1892; the following classification is taken from that volume:—

LIAS OF YORKSHIRE.

UPPER LIAS.	*Striatulus* beds	Zone of *Ammonites jurensis*.
	Alum shale	,, *A. communis*.
	Jet shales	,, *A. serpentinus*.
	Grey shales	,, *A. annulatus*.
MIDDLE LIAS.	Ironstone series	,, *A. spinatus*.
	Sandy series	,, *A. margaritatus*.
LOWER LIAS.	Upper series with Ironstone nodules	,, *A. capricornus*.
		,, *A. Jamesoni*.
		,, *A. armatus*.
		,, *A. oxynotus*.
	Lower series with sandy and marly bands	,, *A. Bucklandi*.
		,, *A. angulatus*.
		,, *A. planorbis*.

[1] Tate & Blake (76).
[2] Witham (33).
[3] Fox-Strangways (92).

Reference must be made also to Horace B. Woodward's important memoir on the Lias of England and Wales (Yorkshire excepted). He gives the following list of plants from Dorsetshire, Yorkshire, and other parts of England :—[1]

Brachyphyllum solitarium, Phill. Warwickshire. (Lower Lias.)	
Pachyphyllum peregrinum, Sternb. Dorsetshire, Gloucestershire, and Yorkshire. (Zones of *A. Bucklandi*, *A. oxynotus*, *A. serpentinus*.)	= *Pagiophyllum peregrinum* (Sternb.).
Cycadites concentricus, Rich. This specific name was proposed by Richards for a Lias specimen in the British Museum.	= *Cycadites rectangularis*, Brauns.
C. rectangularis, Brauns. Dorsetshire. (Lower Lias.)	= *C. rectangularis*, Brauns.
Cycadeoidea pygmæa, L. & H. Dorsetshire. (Lower Lias.)	= *Cycadeoidea pygmæa*, L. & H.
Otozamites acuminatus, L. & H. Somersetshire and Warwickshire. (Lower Lias.)	= ? *Otozamites obtusus* (L. & H.). I have not seen any example of *O. acuminatus* from English Liassic beds.
O. gracilis, Phill. Gloucestershire. (Zone of *A. planorbis*.)	This species of Phillips is, I believe, identical with *Williamsonia pecten* (Phill.),[2] a type not represented among Liassic Cycads.
O. obtusus, L. & H. Devonshire, Dorsetshire, Somersetshire, Gloucestershire, Warwickshire, and Yorkshire. (Zones of *A. planorbis* and *A. Bucklandi*.)	= *Otozamites obtusus* (L. & H.).
Ptilozamites Bergeri, Göpp. Dorsetshire. (Lower Lias.)	= *Ctenopteris cycadea*, Brongn.
P. Leckenbyi, Bean. Dorsetshire and Yorkshire. (Lower Lias.)	= *Ctenopteris cycadea*, Brongn.
Yatesia gracilis, Carr. Dorsetshire.	= *Cycadeoidea gracilis* (Carr.).
Clathropteris platyphylla (Göpp.). Gloucestershire. (Lower Lias.)	= *Clathropteris platyphylla* (Göpp.). (Described under Rhætic plants in the present volume, p. 18.)

[1] Woodward, H. B. (93).

[2] Seward (00), p. 190.

Lomatopteris jurensis, Kurr. Warwickshire. (Zone of *A. planorbis*.)	= *Thinnfeldia rhomboidalis*, Ett.
Macrotæniopteris asplenioides, Ett. Dorsetshire. (Lower Lias.) *Pachypteris*, sp. Dorsetshire. (Lower Lias.)	I have not seen any specimens which conform to the characters of these genera.
Equisetites Brodiei, Buck. Gloucestershire and Worcestershire. (Lower Lias.)	= *Equisetites Muensteri*. (Described under the Rhætic plants, p. 12.)
Chara liassica, Moore. Somersetshire.	Mr. H. B. Woodward notes that the horizon is doubtful.
Chondrites bollensis, Quen. Somersetshire and Oxfordshire. (Zone of *A. serpentinus*.) *Phymatoderma liassicum*, Schloth. Dorsetshire. (Lower Lias.) Fucoids. Somersetshire, Gloucestershire, and Northamptonshire.	Valueless as records of Algæ.

The majority of the Liassic plants have been found in Dorsetshire and East Yorkshire, but, as noted in the above list, a few species are recorded from other counties. The species of *Chara* mentioned in Woodward's list as of doubtful origin was recorded by Moore [1] from Charterhouse; he mentions a single 'seed-vessel,' but does not publish a drawing.

For the sake of convenience all the Liassic species are dealt with together; they are arranged botanically and not geographically.

Class THALLOPHYTA.

The Museum collection does not include any specimens that are worthy of record as fossil Algæ. In their work on the Lias of the Yorkshire Coast, Tate & Blake institute two names for fossils which they regard as algæ. The type-specimens of their two 'species,' *Chordophyllites cicatricosus* and *Nulliporites furcillatus*,[2] which are in the Jermyn Street Museum, should not, in my opinion, be included in any list of undoubted plants. The cast named

[1] Moore (67), p. 538.
[2] Tate & Blake (76), pp. 156, 233, pl. xiv. figs. 7, 9.

Chordophyllites represents what appears to be a stem with irregular surface ridges or wrinklings, and gives one the impression of a fragment of some soft plant which has undergone shrinkage.

The other specimen, *Nulliporites*, may be of animal origin, and cannot be assigned with confidence to the Algæ. In the Whitby Museum there is another specimen of *Chordophyllites* identified by Tate and obtained from the *Ammonites spinatus* beds at Staithes.

The two species, *Chondrites bollensis* and *Phymatoderma liassicum*, included in Woodward's list, are not represented in the Museum collection; the names were no doubt applied to specimens which belong to that large class of fossils assigned by some writers to Algæ and by others attributed to animal agency. They may be regarded as valueless from a botanical point of view. The doubtful origin of Moore's *Chara*, already mentioned, prevents us accepting the record as trustworthy evidence of a Liassic species of Charophyta.

Class EQUISETALES.

Genus **EQUISETITES.**

? **Equisetites Muensteri**, Sternberg.

V. 2092. A portion of a crushed stem, possibly referable to *Equisetites Muensteri*, Sternb., but too indistinct to determine with confidence.

Lyme Regis. *Purchased*, 1889.

Class FILICALES.

Genus **THINNFELDIA**, Ettingshausen.

[Abh. k.k. geol. Reichs. Bd. i. Abth. 3, No. 3, p. 2, 1852.]

This generic name was instituted by Ettingshausen for some specimens from the Lias of Steierdorf, and defined as follows :—

"Rami teretes vel subalati. Folia disticha, alterna oppositave, rhomboidea, ovalia vel lanceolata vel linearia, flabellatim vel pinnatim venosa."

In 1853 Andrae[1] substituted Brongniart's genus *Pachypteris* for *Thinnfeldia*, and spoke of Ettingshausen's type-species as *Pachypteris thinnfeldia*. The genus *Pachypteris* is defined by Brongniart as being characterised by entire pinnules, which are either without veins or provided with a single vein. In some large and well-preserved specimens figured by Zigno[2] from Italian Jurassic rocks, which are no doubt generically identical with *Pachypteris*, the pinnules are traversed by several divergent veins. Species of *Pachypteris* or *Dichopteris*, in which the ultimate segments possess spreading and forked veins, bear a marked resemblance to *Thinnfeldia*; it is probable that the two genera are very closely allied. In 1867 Schenk[3] described examples of *Thinnfeldia* from Rhætic beds of Franconia, and published drawings of the epidermal cells and stomata; he placed the genus among the Cycads. The most satisfactory evidence so far adduced as to the systematic position of *Thinnfeldia* has been furnished by Raciborski[4] in his description and illustration of a specimen from the Jurassic rocks of Poland. This author gives figures of a leaflet bearing the impressions of what appears to be a circular sorus showing the boundaries of individual sporangia. A specimen of the genus in the British Museum (V. 5950) shows two rows of contiguous polygonal prominences, one row on either side of the midrib, which agree in size and shape with the sori of recent ferns; but until we obtain sporangia we cannot speak with certainty as to the precise nature of the fertile frond.

In all probability *Thinnfeldia* is a fern, but we are unable, through lack of good fertile specimens, to determine its family position. The Rhætic fronds from Scania described by Nathorst as species of his genus *Ptilozamites*[5] no doubt belong to plants closely allied to *Thinnfeldia*. Some American Permian specimens recently referred by Sellards[6] to a new genus *Glenopteris* might, I think, be fairly included in *Thinnfeldia*.

---

[1] Andrae (53), p. 43.
[2] Zigno (56), pls. xii., xiii., xv.
[3] Schenk (67), p. 105, pl. xxvii.
[4] Raciborski (94), pl. xx. figs. 1–2.
[5] Nathorst (78), pls. xii. etc.
[6] Sellards (00), pl. xxxix.

There is considerable confusion in regard to the terminology applied to various Rhætic and Jurassic fronds, or portions of fronds, which bear a close resemblance to Ettingshausen's *Thinnfeldia rhomboidalis*, the type species of the genus. The two best known and widely distributed species of this genus are *T. rhomboidalis*, with which may be included several forms designated by authors as distinct species, and *T. odontopteroides* (Morr.). The latter type, originally described by Morris in Strzelecki's book on New South Wales as *Pecopteris odontopteroides*,[1] is characteristic of Rhætic rocks in South America, South Africa, and Australia. The examination of several specimens of this species from the Stormberg beds of the Cape has convinced me that the distinction between Morris's species and Ettingshausen's European species is not always well defined. There is, moreover, a striking agreement between some of the specimens described by Nathorst from the Rhætic of Scania as species of *Ptilozamites*[2] and the southern species of *Thinnfeldia*. I have no doubt that *Ptilozamites* and *Thinnfeldia* represent closely allied genera, which, indeed, may not be generically distinct. Another designation that has been employed in the determination of fronds identical in habit with *Thinnfeldia* is *Lomatopteris*. This name was instituted by Schimper[3] for a plant previously described by Kurr as *Odontopteris* (?) *jurensis*,[4] and by Quenstedt as *Neuropteris limbata*. In 1861 Zigno instituted the term *Cycadopteris* for certain Italian specimens which were afterwards included by Schimper under his genus *Lomatopteris*. Saporta,[5] Zeiller,[6] and others have retained both *Lomatopteris* and *Cycadopteris* in addition to *Thinnfeldia*, and the feature noticed by these authors as distinctive of *Cycadopteris* is the presence of a thicker border in the lamina of the ultimate segments. This characteristic margin is well shown in the Dorsetshire specimens represented on Plate IV.; it is not formed by the overfolding of the edge of the pinnule, but is simply a darker (cuticularised) border. The difficulty,

---

1 Seward (03), p. 52.
2 Nathorst (78).
3 Schimper (69), p. 472.
4 Kurr (45).
5 Saporta (73).
6 Zeiller (00), pp. 93, 98.

indeed the impossibility, of distinguishing between specimens figured by different authors as species of *Thinnfeldia*, *Lomatopteris*, and *Cycadopteris*, becomes apparent on a careful comparison of the published drawings. I am disposed to take the view that the better plan is to employ Ettingshausen's generic name *Thinnfeldia* in a fairly comprehensive sense, including in it Zigno's *Cycadopteris* and Schimper's *Lomatopteris*. Such evidence as we possess points to a close relationship between the plants referred to these three genera, and I believe justifies their inclusion under one name. Until we know something of the reproductive organs we cannot expect to arrive at a satisfactory conclusion as to the precise affinity of these various fossils, and we are unable, in the present state of our knowledge, to point to satisfactory distinguishing features of sufficient importance to serve as generic criteria.

This is not the place to attempt a critical examination of the numerous Mesozoic fossils described as species of *Thinnfeldia*, and until we know more about the reproductive organs of the plants which conform more or less closely to Ettingshausen's genus, there is little to be gained by discussing specific distinctions. Attention should, however, be called to a paper published in the Bulletin of the Torrey Botanical Club (August, 1903) entitled *The American Species referred to Thinnfeldia*.[1] The author deals with forms recorded by Fontaine, Hollick, Knowlton, Lesquereux, and Newberry. He comes to the conclusion that many of the species included by American writers under *Thinnfeldia* differ generically from European forms obtained from lower geological horizons. The Middle and Upper Cretaceous species he designates by a new name, *Protophyllocladus*. It is undoubtedly true that the term *Thinnfeldia* has been used in too wide a sense, including plant-fragments which afford no evidence of close relationship. Mr. Berry draws attention to the fact that the later Mesozoic forms from American rocks referred to *Thinnfeldia* do not closely conform to the typical species recorded from Jurassic and Rhætic strata. He states that a careful comparison of the fossils from Cretaceous rocks with the living material available at the New York Botanical Garden, has "thoroughly convinced" him that "all the Middle and

---

[1] Berry (03).

late Cretaceous species should be included in the coniferous family Taxaceæ." He adds—"They may perhaps form a link between the Podocarpeæ and the Taxeæ, and while they are unmistakably related to *Phyllocladus*, their extremely large size compared with the existing members of that genus and other minor differences, such as petiolate forms and the general absence of crenate margins, make it desirable to refer them to a new genus, which, while indicating their proper relationship, shall keep them distinct from the small-leaved species." I maintain that this conclusion is not supported by satisfactory evidence. Before accepting the above statement it is important that we should inquire into the nature of the evidence on which it is based. My view is that the use of the generic name *Protophyllocladus* is a retrograde step which should be strongly deprecated. Palæobotanical literature contains too many designations characterised by the prefix 'Proto-' which cannot be defended by evidence which is likely to be accepted by critical systematists. Mr. Berry constantly uses the word 'leaf' in speaking of the fossil specimens and in his references to recent species of *Phyllocladus*, thus ignoring the fact that the assimilating organs of the New Zealand Conifer are branches (phylloclades), while the true leaves assume the form of small inconspicuous scales. Any facts that can be adduced bearing on the past history of so isolated a type as the New Zealand genus *Phyllocladus* would be particularly welcome; but to describe the American Cretaceous fossils as representatives or near relations of that genus can tend only to increase the distrust on the part of botanists of palæobotanical records. We know nothing of the reproductive organs of the Cretaceous plants referred to *Protophyllocladus*, nor, so far as I am aware, have we any justification for assuming that the superficial resemblance of the apparent leaf-fragments to the phylloclades of *Phyllocladus* amounts to morphological identity.

Thinnfeldia rhomboidalis, Ettingshausen.

[Abh. k.k. geol. Reichs. Bd. i. Abth. 3, No. 3, p. 2, pl. i. figs. 4–7, 1852.]

(Pl. IV. Figs. 1–3.)

1845. ? *Odontopteris jurensis*, Kurr, Beit. foss. Flor. Jura Württembergs, p. 12, pl. ii. fig. 1.

1852. *Thinnfeldia rhomboidalis*, Ettingshausen, Abh. k.k. geol. Reichs. Bd. i. Abth. 3, No. 3, p. 2, pl. i. figs. 4–7.

1856. *Cycadopteris Brauniana*, Zigno, Flor. foss. Oolit. vol. i. p. 155, pls. xvi.–xvii.
C. heterophylla, ibid. pl. xviii.
1858. ? *Odontopteris jurensis*, Quenstedt, Der Jura, p. 800, pl. xcix. fig. 9.
1867. *Thinnfeldia rhomboidalis*, Schenk, Foss. Flor. Grenz. p. 116, pl. xxvii. figs. 1–8.
T. decurrens, ibid. p. 114, pl. xxvi. figs. 1–5.
T. obtusa, ibid. p. 115, pl. xxvi, figs. 6–8.
1869. ? *Lomatopteris jurensis*, Schimper, Trait. pal. vég. vol. i. p. 472.
Thinnfeldia decurrens, ibid. p. 495.
T. rhomboidalis, ibid. p. 496.
1873. *Cycadopteris Brauniana*, Saporta, Pal. Franç. vol. i. p. 421, pl. liv. fig. 5; pl. lvii. figs. 3–4; pl. lviii. figs. 1–5.
Thinnfeldia rhomboidalis, ibid. p. 343, pl. xliii.
Lomatopteris jurensis, ibid. pl. lv. figs. 1–4.
? *Thinnfeldia incisa*, ibid. p. 348, pls. xl., xlii.
1880. Cf. *Lomatopteris liasina*, Morière, Bull. Soc. Linn. Normandie, vol. iv. [3], p. 361, pl. iv.
Pachypteris (*Thinnfeldia*) cf. *decurrens*, Schenk, and *rhomboidalis*, Ett., Nathorst, Öfver. k. Vet.-Akad. Förh. 1880, p. 84.
1888. Cf. *Thinnfeldia lancifolia*, Szajnocha, Sitzb. k. Akad. Wiss. Wien. vol. xcviii. Abth. 1, p. 13, pl. i. figs. 4–7.
1893. *Lomatopteris jurensis*, Woodward, Lias, p. 378.
1894. *Thinnfeldia rhomboidalis*, Raciborski, Flor. Kopalna, pl. xix. figs. 9–15.
1903. *T. rhomboidalis*, Seward, Foss. Flor. Cape Colony, p. 57, pl. viii. fig. 1.

Frond pinnate or bipinnate; in the larger examples the pinnate leaves show a tendency in the lobing of the segments to assume a bipinnate habit. Rachis stout, bearing alternately disposed segments attached laterally. Pinnules leathery, varying considerably in size and shape; those near the base of a frond may be short and more or less ovate in form, while others are linear and reach a length of rather more than 3 cm.; apices of pinnules bluntly rounded, upper margin of the lamina curves downwards towards the rachis, lower margin decurrent; midrib well marked in the longer segments, but in the smaller ones it dies out a short distance from the base or (Text-fig. 2) is entirely absent; secondary dichotomously branched veins spring from the midrib at a very oblique angle.

Ettingshausen defined his species *Thinnfeldia rhomboidalis*, founded on material from the Lias of Steierdorf, as follows:—

"Th. ramis elongatis, subalatis, striatis, foliis coriaceis, rigidis, rhomboideis vel ovalibus, obtusis v. acutis, 12–20 mm. longis,

5–12 mm. latis, integerrimis, obliquis, basi coarctata et decurrenti sessilibus, approximatis, distichis, flabellatim, venosis, nervis tenuibus, subrectis; simplicibus vel dichotomis, marginem versus divergentibus."

The specimens from the Lower Lias of Dorsetshire, as represented in Pl. IV. Figs. 1–3, exhibit a considerable range in regard to the size and form of the ultimate segments. The specimen figured by Kurr as *Odontopteris jurensis* may perhaps be specifically identical with the English form; if we imagine the marginal undulations of the pinnules shown in Pl. IV. Fig. 3 extended into distinct lobes, we should have a frond like that figured by Kurr. Although it is not unlikely that the difference between *Odontopteris jurensis* as figured by Kurr and the English specimens does not amount to specific difference, I prefer to adopt Ettingshausen's specific designation *rhomboidalis*, as the specimens he figures are more nearly identical in form with those from Lyme Regis. The examples of *Thinnfeldia* figured by Schenk from the Rhætic of Franconia as *T. decurrens* and *T. obtusa* differ in so slight a degree from *T. rhomboidalis* that we may regard them as constituting one type. Zigno's Italian examples referred by him to *Cycadopteris Brauniana* can hardly be specifically distinguished from *Thinnfeldia rhomboidalis.*

The type of plant represented by the fronds classed under *T. rhomboidalis* had an extended geographical range during the Rhætic and Lower Jurassic periods. It is hopeless to attempt to define within narrow limits the specific types of *Thinnfeldia* fronds; we find numerous fragments of leaves in various parts of the world exhibiting a close resemblance with one another and differing in what we may fairly consider unimportant details. The range of *Thinnfeldia* extended into the southern hemisphere, where the species *T. odontopteroides* is particularly abundant, in South African Rhætic beds, in Australia and South America;[1] in India[2] also a form has been found which may be identical with the European type, *T. rhomboidalis.* Nathorst, in his interesting notes on fossil plants in British museums, noted the very close agreement

[1] Solms-Laubach & Steinmann (99), pl. xiv. fig. 2.
[2] Feistmantel (77), pls. xxxix. and xlvi.

between the Lyme Regis leaves and those described by Schenk and Ettingshausen from Central Europe. The specimens from Australia referred by Tenison-Woods[1] to *Thinnfeldia falcata* and *Gleichenia lineata* probably represent forms closely allied to *T. rhomboidalis*.

52,751*a*. Pl. IV. Fig 1.

A fairly complete frond, 12·5 cm. long; the stout rachis bears crowded, broadly linear segments, showing traces of a midrib in the lower part of the lamina and oblique secondary veins. The edges of the pinnules show a dark border which suggests a folding over of the lamina, but a microscopical examination of the leaf-substance affords no evidence of a revolute edge. The broad rounded tip of the frond is well preserved; the entire lamina at the apex passes gradually through a lateral lamina with confluent lobes into distinct linear segments.

Lyme Regis. *Purchased*, 1874.

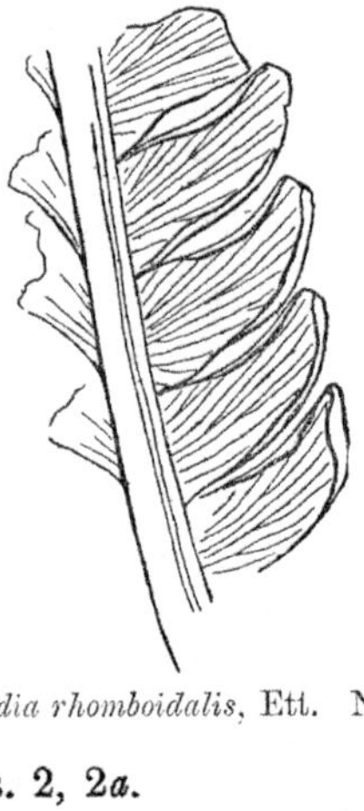

FIG. 2.—*Thinnfeldia rhomboidalis*, Ett. No. V. 40,674. ×1½.

52,672. Pl. IV. Figs. 2, 2*a*.

The apical portion of a frond of which the pinnules (Fig. 2*a*) show very well preserved, dichotomously branched, secondary veins, passing obliquely into a midrib. The dark border of the segments is clearly marked, and the surface of the lamina shows numerous small papillæ.

Lyme Regis. *Purchased*, 1870.

[1] Tenison-Woods (83).

52,751. Pl. IV. Fig. 3.

The specimen, of which a portion is represented in the figure, is 15 cm. in length. The longer segments reach a length of 3 cm.; the lamina of some of them has a slightly lobed margin.[1] The upper edge of the pinnule bends sharply downwards at the base, forming a deep sinus, while the lower margin of the lamina is decurrent on the rachis.

Lyme Regis. *Purchased*, 1874.

40,674. Text-fig. 2 (one and a half times nat. size). In this smaller frond there is no midrib in the pinnules, the branched veins spreading from the base of each pinnule.

Lyme Regis. *Purchased*, 1859.

V. 17. A specimen 10 cm. long, with a petiole 1·7 cm. in length, terminating in a slightly swollen base.

Lyme Regis. *Purchased*, 1881.

Other specimens:—**35,046, 38,350, 38,351, 52,510, 52,858.**

Lyme Regis. *Purchased*, 1856–76.

PLANTA INCERTÆ SEDIS.

Genus **CTENOPTERIS**, Saporta, *ex* MS. Brongniart.

[Saporta, Pal. Franç. vol. i. p. 351, 1873.]

In my former volume[2] on Jurassic plants I adopted Nathorst's generic name *Ptilozamites* for the Inferior Oolite species originally named by Zigno *Odontopteris Leckenbyi*, and afterwards referred by Leckenby to the genus *Ctenis* and by Nathorst to *Ptilozamites*. The designation *Ctenozamites* of Nathorst was added as a subgeneric title for bipinnate, as distinct from pinnate, fronds of the *Ptilozamites* type.

It is a matter of secondary importance whether we adopt the names *Ptilozamites* or *Ctenozamites* of Nathorst, or revert to the older genus *Ctenopteris*. Professor Zeiller, in his admirable *Éléments de Paléobotanique*, makes use of *Ctenopteris*, and it may

---

[1] Compare Zigno (56), pls. xvii.–xviii.

[2] Seward (00), p. 237.

tend to greater uniformity in nomenclature if we give precedence to this genus. The generic name *Ctenopteris* was first used by Saporta, at the suggestion of Brongniart, for a Jurassic species, *Ctenopteris cycadea*; it is defined as follows:—"Frons pinnata vel bi-tripinnata, pinnæ elongato-lineares pinnati-partitæ basi exappendiculatæ, pinnulæ basi tota adnatæ decurrentes inter se liberæ versus apicem pinnarum plus minusve confluentes, nervi omnes costa exorientes simplices furcatique divergentes, nervo medio nullo, nervulis mediis dense quandoque fasciculatis; fructificatio ignota."

Ctenopteris cycadea (Brongn.).

[Hist. vég. foss. p. 387, pl. cxxix. figs. 2–3, 1828.]

(Pl. III. Figs. 1, 1*a*.)

1828. *Filicites cycadea*, Brongniart, Hist. vég. foss. p. 387, pl. cxxix. figs. 2–3.
1832. *Odontopteris cycadea*, Berger, Verstein. Coburg Geg. p. 23, pl. iii. figs. 2–3.
1836. *O. Bergeri*, Göppert, Syst. fil. foss. p. 219.
1869. *Cycadopteris Bergeri*, Schimper, Trait. pal. vég. vol. i. p. 487.
1873. *Ctenopteris cycadea*, Saporta, Pal. Franç. vol. i. p. 355, pl. xl. figs. 2–5; pl. xli. figs. 1–2.
1880. *C. cycadea*, Nathorst, Öfvers. k. Vet. Akad. Förh. 1880, p. 45.
1884. *Ptilozamites Bergeri*, Richards, Synopsis Foss. Cycad. p. 2.
1887. *Ctenozamites cycadea*, Schenk, Foss. Pflanz. Albourskette, p. 5, pls. iii.–iv., vi.–ix.
1891. *Ctenopteris cycadea*, Saporta, Pal. Franç. vol. iv. pl. lxvi.
1893. *Ptilozamites Bergeri*, Woodward, Lias, p. 378.
1900. *Ctenopteris cycadea*, Zeiller, Eléments Paléobot. p. 101, fig. 75.

Frond bipinnate; pinnæ attached to the broad rachis at an acute angle, bearing broad, linear, entire pinnules; pinnules slightly falcate, entire, attached to the axis of the pinna by the whole of the base; no midrib, the lamina of the ultimate segments traversed by several slightly divergent and dichotomously branched veins.

The Inferior Oolite species, previously described as *Ptilozamites* (*Ctenozamites*) *Leckenbyi*,[1] bears a close resemblance to *Ctenopteris*

---

[1] Leckenby (64), pl. x. fig. 1.

cycadea. Portions of a large frond recently figured by Zeiller from the Rhætic of Tonkin as *Ctenopteris Sarraini*, Zeill.,[1] represent another example of the genus.

40,674. Pl. III. Figs. 1, 1*a*.

Part of a bipinnate frond; a broad and flattened rachis gives off pinnate branches bearing ultimate segments attached by the whole of the base and traversed by several forked and slightly spreading veins. The pinnules are approximately 1·8 cm. long and 8 mm. broad.

Lyme Regis. *Purchased*, 1859.

Class CYCADOPHYTA.

This class-designation recently proposed by Nathorst[2] serves a useful purpose, and forms a welcome addition to botanical nomenclature. In a great number of cases we are without the means of referring fossil Cycadean fronds either to the Bennettiteæ or to the Cycadaceæ, the chief difference between the two sections being afforded by the morphological characters of the reproductive organs. It is, therefore, a convenience to make use of a designation which includes both the true Cycads and the extinct Bennettitaceæ.

Genus **CYCADITES**, Sternberg.

[Flor. Vorwelt, iv. p. xxxii. 1825.]

Cycadites rectangularis, Brauns.

[Palæontogr. Bd. ix. p. 56, pl. xiv. fig. 1, 1862.]

(Pl. IV. Figs. 4, 4*a*; Pl. II. Fig. 1; Text-fig. 3.)

1832. *Cycadites pectinatus*, Berger, Verstein. Coburg, p. 23, pl. iii. fig. 4.
1862. *C. rectangularis*, Brauns, Palæontogr. vol. ix. p. 56, pl. xiv. fig. 1.
1863. Cf. *C. rajmahalensis*, Oldham & Morris, Gondwana Flor. pls. vii.-viii.
Cf. *C. conferta*, ibid.
1867. *C. rectangularis*, Schenk, Flor. Grenz. p. 157, pl. xxxv. fig. 11.
1870. *C. rectangularis*, Schimper, Trait. pal. vég. vol. ii. p. 178.
C. pectinatus, ibid. p. 178.

---

[1] Zeiller (02), pls. vi.–viii.

[2] Nathorst (02), p. 3.

1875. *C. rectangularis*, Saporta, Pal. Franç. vol. ii. p. 69, pl. lxxxiii. figs. 1–4.
Cf. *C. Delessei*, ibid. p. 73, figs. 5–7.
1884. *C. concentricus*, Richards, Synopsis Foss. Cycad. p. 1.
1893. *C. concentricus*, Woodward, Lias, p. 378.
C. rectangularis, ibid. p. 378.
1895. *C. rectangularis*, Lignier, Mém. Soc. Linn. Normandie, vol. xviii. p. 27, pl. vii. fig. 19.
1902. Cf. *C. Saladini*, Zeiller, Flor. foss. Tonkin, pl. xli.

Frond pinnate; rachis broad, bearing narrow linear and crowded pinnæ; pinnæ attached laterally by a broad base, of uniform breadth, terminating in an acuminate tip, they may be either at right angles to the axis of the frond or curved slightly upwards; the lamina is traversed by a well-defined midrib.

The genus *Cycadites*[1] is less abundantly represented in Mesozoic floras than the comparatively frequent use of the name implies. Sternberg's generic name was formerly used as a general designation of Cycadean fronds which have since been separated into several distinct genera. Other leaves have been referred incorrectly to *Cycadites* instead of to the genus *Dioonites*; if, as occasionally happens in fossil as in recent Cycadean fronds, the lamina of the pinnæ has a revolute margin, an impression or cast presents an appearance suggestive of a midrib. Among existing Cycads the species *Encephalartos Ghellinckii*, Lem.,[2] affords a good example of a frond that might be easily mistaken for a *Cycadites* if preserved as a fossil.

Cycadites rectangularis has a fairly wide distribution in Rhætic and Lower Jurassic strata, and represents a typical example of the genus. Unusually large and well-preserved Wealden specimens of *Cycadites* have been described from England and Portugal respectively as *Cycadites Saportæ*, Sew.,[3] and *C. tenuisectus*, Sap.[4]; in all probability these two specific names stand for one species.

52,665. Pl. IV. Figs. 4, 4*a*.

Part of a frond 20·5 cm. in length. The rachis is 1 cm. broad, and bears numerous linear pinnæ, none of which are complete;

---

[1] Solms-Laubach (91), p. 86; Seward (95), p. 23.
[2] Seward (98[2]).
[3] Seward (95), p. 29, pls. iii., vi., viii.
[4] Saporta (94), p. 171, pl. xxxii.

the longest is 5 cm. long and 2 mm. broad. The frond is seen from below, so that the pinnæ are cut off abruptly by the edges of the rachis; each pinna appears to be traversed by a single vein.

Lyme Regis. *Purchased*, 1870.

V. 4076. Text-fig. 3.

A specimen 8 cm. in length, of which a portion is represented in the figure; the pinnæ are straighter and attached to the rachis at a wider angle than in the example shown in Fig. 4, Pl. IV.

Lyme Regis.

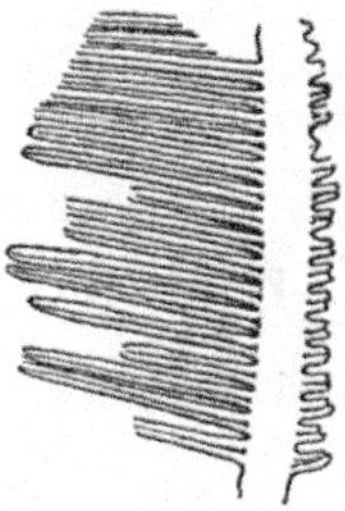

FIG. 3.—*Cycadites rectangularis*, Brauns. V. 4076. Nat. size.

? Cycadites rectangularis.

V. 9009. Pl. II. Fig. 1.

This frond was obtained at Wansford, in Northamptonshire, by C. W. Peach. The rachis has a length of 27 cm., and the pinnæ, which are attached to the axis at right angles, have an average length of 4 cm. and a breadth of from 2 to 3 mm. In some of the pinnæ there appear to be indications of a midrib, but the preservation is not sufficiently good to enable one to state confidently that the frond is a true *Cycadites*.

Wansford.

Genus **OTOZAMITES,** Braun.

[Braun, in Münster, Beit. Petrefact. Heft vi. p. 36, 1843.]

Otozamites obtusus (Lindley & Hutton).

[Foss. Flor. pl. cxxviii. 1834.]

(Pl. I. Figs. 1, 3, 5.)

1824. "Fossil Fern," De la Beche, Trans. Geol. Soc. vol. i. [2], p. 45, pl. vii. figs. 2-3.

1825. *Filicites Bechei*, Brongniart, Ann. Sci. nat. vol. iv. p. 422, pl. xix. fig. 4.

1825. *F. Bucklandi*, Brongniart, ibid. pl. xix. fig. 3.
1828. *Zamites Bechei*, Brongniart, Prodrome, p. 94.
Z. Bucklandi, ibid.
1834. *Otopteris obtusa*, Lindley & Hutton, Foss. Flor. pl. cxxviii.
1836. *Odontopteris otopteris*, Göppert, Foss. Farrn. p. 211.
1838. *Cyclopteris obtusa*, Sternberg, Flor. Vorwelt, vii. p. 133.
Odontopteris Bucklandi, ibid. p. 79.
1842. *Zamites Bechei*, Miquel, Monog. Cycad. p. 38.
1843. *Z. Bechei*, Göppert, Foss. Cycad. p. 43.
Z. Bucklandi, ibid.
1845. *Odontopteris otopteris*, Unger, Synopsis, p. 51.
Zamites Bechei, ibid. p. 152.
Z. Bucklandi, ibid. p. 152.
1847. *Odontopteris otopteris*, Unger, Chlor. Prot. p. xxxvii.
Zamites Bechei, ibid. p. lxii.
Z. Bucklandi, ibid.
1848. *Odontopteris otopteris*, Bronn, Ind. Pal. p. 838.
Zamites Bechei, ibid. p. 1377.
Z. Bucklandi, ibid. p. 1377.
1849. *Otozamites obtusus*, Brongniart, Tableau, p. 104.
O. Bechei, ibid.
O. Bucklandi, ibid.
1850. *Odontopteris otopteris*, Unger, Gen. spec. p. 89.
Zamites Bechei, ibid. p. 283.
Z. Bucklandi, ibid. p. 284.
1851. *Otozamites obtusus*, Miquel, Over Rangs. foss. Cyc. p. 210.
O. Bechei, ibid.
O. Bucklandi, ibid.
1854. *Palæozamia Bechei*, Morris, Cat. Brit. Foss. p. 15.
P. Bucklandi, ibid.
1856. *Otozamites Bechei*, Bornemann, Organ. Rest. Lettenkohl. p. 52.
1870. *O. obtusus*, Schimper, Trait. pal. vég. vol. ii. p. 171.
O. Brongniarti, ibid. p. 172.
1875. *O. Brongniarti*, Saporta, Pal. Franç. vol. ii. pl. ciii. fig. 4.
O. Hennoquei, ibid. p. 143, pls. c.–ci.
1884. *O. brevifolius*, var. *Bucklandi*, Richards, Synopsis, p. 6.
O. Bechei, ibid. p. 7.
1893. *O. obtusus*, Woodward, Lias England and Wales, p. 378.
1901. Cf. *O. Ameghinoi*, Kurtz, Lias Piedra Pintada, p. 14, pl. iii. fig. 6.

Frond pinnate; pinnæ usually more or less falcate, occasionally straight, attached alternately to the upper face of the rachis, upper edge of the base is strongly auriculate; the upper edge of the lamina is either straight or, more usually, slightly concave, while the lower margin is more strongly convex, curving upwards somewhat abruptly to the blunt apex; pinnæ occasionally imbricate (young fronds). Veins divergent in the basal portion of the pinnæ, extending in an oblique direction to the upper and lower edge of the lamina.

Otozamites obtusus, founded by Lindley & Hutton on a specimen from the Lias of Axminster, and now preserved in the Oxford Museum, forms one of several species of the genus which are so closely linked together that it is often impossible to recognise satisfactory distinguishing features.

The type-specimen of *Otozamites obtusus* consists of a portion of a frond 8 cm. long, with very clearly preserved pinnæ.

Otozamites graphicus, a characteristic Inferior Oolite species, represents a type of frond very similar in shape and in the form and venation of the pinnæ to *O. obtusus*. From Jurassic rocks of France, Italy, and elsewhere, numerous examples of *Otozamites* have been described which demonstrate the abundance of this generic type, and the impossibility of defining the limits of the various species. Attention has elsewhere been drawn to the futility of attempting to recognise specific distinctions in the slight variations in shape which are met with in the pinnæ of *Otozamites* fronds, and, to avoid the danger of carrying artificial differences too far, I have suggested the use of the specific name *obtusus* in a comprehensive sense.

Specimens of *Otozamites obtusus* from the Lias of England are fairly common; in addition to the well-preserved fronds in the British and Oxford Museums, examples are included in the collections in the Jermyn Street Museum and in the Museums of Bath, Bristol, and Cardiff.

40,672. Pl. I. Fig. 1.

An almost perfect frond 24 cm. long and 3·1 cm. broad at its widest part. The pinnæ are obtusely pointed, slightly falcate, and have a prominent lobe on the upper edge of the base. Veins spreading from the centre of the base, cutting the edges of the pinnæ obliquely.

Near Lyme Regis. *Purchased*, 1859.

[**1155.** Pl. I. Fig. 2.

A specimen from near Baireuth; the figure represents a portion of a frond 20 cm. in length. This is undoubtedly the species figured by Schenk as *Otopteris Bucklandi*, but it is probably not specifically identical with the English species, *Olozamites obtusus*; the lower edge of the segments in the German type are more gradually curved upwards to the apex, as contrasted with the more sudden bend near the tip of the English species; the difference, however, is very slight, and whether identical or not the two forms belong to the same general type. *Braun Coll.*]

V. 22. Pl. I. Figs. 3, 3*a*.

Two specimens, one the reverse of the other. The venation and auriculate bases (Fig. 3*a*) of the pinnæ are very clearly shown.

Bridgewater, Somerset. *Purchased*, 1881.

39,059. Pl. I. Fig. 5.

A good example illustrating the gradual tapering in breadth towards the base of a frond; probably not fully expanded. Some of the pinnæ are clearly falcate, while others may be described as straight.

Langport, Somerset. *Presented by Mr. E. W. Bagehot.*

V. 3362. Part of a frond 12 cm. long; pinnæ contiguous, similar to those of 39,059 (Fig. 5).

Lower Lias, Worcestershire. *Brodie Coll.*

V. 3456. Part of a leaf 4·2 cm. broad; pinnæ contiguous.

Lower Lias, Gloucestershire. *Buckman Coll.*

Other specimens:—**47,045** (very imperfect specimen, 36 cm. long, from Lyme), **52,668, 52,751.** *Purchased*, 1866–74.

CYCADEAN STEMS.

The nomenclature and generic separation of fossil Cycadean stems present considerable difficulties. In dealing with the enormous number of trunks discovered in recent years in Jurassic and Cretaceous rocks of the United States,[1] Professor Lester Ward[2] has adopted the plan of using Buckland's generic name *Cycadeoidea* in a wide sense, including species possessing the characters of Carruthers' genus *Bennettites*. In the monograph on Italian Cycad stems by Capellini & Solms-Laubach[3] *Cycadeoidea* is also employed

[1] Ward (95), (00), (00[2]).

[2] Ward (94).

[3] Capellini & Solms-Laubach (91).

in a similar comprehensive sense, but these authors retain the designation *Bennettites* for the species *B. Gibsonianus*, as we possess a fairly complete knowledge of the morphology of its female reproductive organs. Since Carruthers' *Bennettites* was first described[1] several other examples of the same genus have been discovered[2] in a state of preservation that admits of minute investigation of the fertile shoots. There can be little doubt that the majority of Mesozoic Cycads belong to the section Bennettitales, and bore reproductive organs of the type represented by *Bennettites Gibsonianus.* In an earlier volume[3] I expressed the opinion that we should do more wisely to restrict the generic name *Bennettites* to stems bearing the bennettitean inflorescence; the lateral reproductive shoots are usually recognised without difficulty, and we may safely assume, in cases where special lateral branches are present, that they bore flowers similar to those of *Bennettites Gibsonianus.* There are, however, various forms of trunks which present no signs of lateral fertile shoots; they bear a close resemblance to the common type of recent Cycadean stem in being covered with a thick envelope or armour of persistent petiole-bases.

Stems of this category have been described as species of *Cycadeoidea*, *Bucklandia*, *Yatesia*, and other genera. Lester Ward, in a paper dealing with *A Revision of the genus Cycadeoidea of Buckland*,[4] refers to the persistent character of the leaf-bases in *Yatesia*, *Fittonia*, and *Platylepis*, as a feature distinguishing them from *Cycadeoidea.* This distinction is, however, hardly a valid one. The stem of *Cycadeoidea*, using the term in the sense in which it is employed by Ward and other authors, is covered with rhomboidal areas, sometimes excavated by decay into hollow spaces bounded by a silicified mass of chaffy ramenta, or presenting the appearance of flat or convex areas; these areas are the bases of petioles from which the rest of the frond has been cut off by means of a definite absciss-layer. In *Yatesia*, *Fittonia*, and other genera, the leaf-bases may be in some cases more prominent, but I am unable to recognise

---

[1] Carruthers (70).

[2] For more recent descriptions of *Bennettites*, see Solms-Laubach (91^2); Lignier (94), (03); Scott (00), p. 445; Coulter & Chamberlain (01); Wieland (99), (99^2).

[3] Seward (95), p. 134.

[4] Ward (94).

any essential distinction between these stems and those of *Cycadeoidea*. We cannot enter into a detailed consideration of the various other generic terms applied to Cycadean stems, but mention may be made of one recent addition by Lester Ward. He has instituted the genus *Cycadella*[1] as a generic name for Cycadean stems especially characterised by the abundance of the woolly or scaly ramenta which formed a thick covering over the greater part of the surface of the stem. It is, I think, doubtful whether this new term is really needed; the extent to which the ramenta may be developed on the petioles and scale-leaves is hardly a character of generic rank, but in all probability the peculiarity of the stems referred to *Cycadella* is an expression of certain external conditions which led to an unusual development of ramental scales. It is hopeless to attempt the separation of Cycadean stems into distinct genera if we take as our criteria such characters as the form of the leaf-bases and their degree of prominence, the relative length of the stem, and other characters of the same kind. The English trunks referred by Carruthers to Presl's genus *Bucklandia* and to the genus *Yatesia* are not distinguished by any striking or important features, and the same criticism may be applied to other forms referred to distinct genera.

Among existing Cycads there are some, e.g. species of *Zamia*, which possess stems covered by a corky investment which shows no signs of persistent leaf-bases.[2] It is probable that stems of this type are also represented among fossil Cycads.

The generic name *Benstedtia*[3] was suggested for some English stems of Lower Greensand age in 1896, and I have more recently made use of the same genus for certain South African stems which are probably Cycadean.

Leaving out of account such stems as there is reason for including in the Cycads and which are without the covering of persistent petiole-bases, it would make for less artificiality in our nomenclature if we grouped under the generic designation *Cycadeoidea* all Cycadean trunks, whether bulbous or of the longer and narrower form, which are covered with persistent leaf-bases, and which show

---

[1] Ward (00), p. 392; (00[2]).

[2] Seward (96), pl. xiv. fig. 1.

[3] Seward (96); (03), p. 34.

no trace of lateral reproductive shoots of the form characteristic of *Bennettites*. The retention of *Bennettites* for stems which bore reproductive shoots like those of *Bennettites Gibsonianus* and several more recently described species, serves to express the fact that we possess some knowledge of the taxonomic position of the plants within the class Cycadophyta.

In an account of a Cycadean stem from the Corallien of l'île de Dives (Vendée), near l'Aiguillon sur Mer, M. Fliche [1] speaks of the generic term *Cycadeoidea* as being applied to stems which probably belong to the Bennettiteæ. This use of the name is, however, not in accordance with the plan now advocated. The designation *Cycadeoidea* may, I consider, be more advantageously applied to Cycadean stems, such as the specimen figured by Fliche, which exhibit no trace of the lateral fertile shoots characteristic of the true Bennettiteæ.

Genus **CYCADEOIDEA**, Buckland.

[Proc. Geol. Soc. London, vol. i. No. 8, pp. 80–81, 1827.]

If used in the comprehensive sense suggested above, but to the exclusion of stems conforming to the genus *Bennettites*, *Cycadeoidea* would include such genera as *Yatesia*, *Bucklandia*, and several others which, so far as we know, differ from one another only in regard to unimportant vegetative features.

Cycadeoidea (Yatesia) gracilis (Carr.).

[Trans. Linn. Soc. vol. xxvi. p. 689, pl. lv. fig. 2, 1870.]

1870. *Yatesia gracilis*, Carruthers, Trans. Linn. Soc. vol. xxvi. p. 689, pl. lv. fig. 2.
1874. *Y. gracilis*, Schimper, Trait. pal. vég. vol. iii. p. 555.
1893. *Y. gracilis*, Woodward, Lias, p. 378.

The following is Carruthers' original diagnosis of this species:—

"Stem cylindrical, very slender, covered with the permanent bases of the petioles, which are imbricated, subrhomboidal, and terminated by a lozenge-shaped cicatrix."

---

[1] Fliche (01), p. 193.

The Jermyn Street Museum collection includes a portion of the bark — consisting chiefly of well-preserved leaf-bases — of a Cycadean stem from Lyme Regis (No. 6397), which is probably specifically identical with the specimens in the British Museum.

V. 4077. Type-specimen of Carruthers (Trans. Linn. Soc. vol. xxvi. pl. lv. fig. 2). The drawing given by Carruthers represents the specimen about one-half the natural size. The fossil, which is 27 cm. long and 4 5 cm. broad, consists of an imperfectly preserved stem partially covered with persistent bases of fronds. These petiole-bases terminate in a flat and well-defined absciss-surface, approximately 7 mm. to 1 cm. in breadth. There is no indication of any alternation of larger and smaller leaf-bases such as characterises the stems of most species of the genus *Cycas*. A somewhat indistinct circular projection on the stem is described by Carruthers as the scar of an axillary organ; this may perhaps mark the position of a lateral branch; a less distinct but similar projection occurs near the base of the specimen.

Lyme Regis. *History unknown.*

52,751. A stem 42 cm. long, but imperfectly preserved. In the lower part the persistent leaf-bases are crowded over the surface, while in the upper portion they are seen in side-view curving upwards from the edge of the central axis, which has a breadth of 1·5 cm. Towards the lower end of the fossil some of the petiole-bases show a clean-cut termination or absciss-surface, about 8 mm. in diameter.

Lyme Regis. *Purchased*, 1874.

52,751*a*. A specimen 16 cm. in length, with an axis nearly 3 cm. wide.

Lyme Regis. *Purchased*, 1874.

? Cycadeoidea pygmæa, Lind. & Hutt.

1835. *Cycadeoidea pygmæa*, Lindley & Hutton, Foss. Flor. pl. cxliii.
1841. *Zamites pygmæus*, Morris, Ann. Mag. vol. vii. p. 116.
1843. *Z. pygmæus*, Göppert, Foss. Cycad. p. 41.
1847. *Cycadeoidea pygmæa*, Unger, Chlor. Prot. p. lxv.
1848. *C. pygmæa*, Bronn, Ind. Pal. p. 370.
1849. *C. pygmæa*, Brongniart, Tableau, p. 104.
Echinostipes pygmæus, Pomel, Vers. Ges. Deutsch. p. 17.

1850. *Cycadeoidea pygmæa*, Unger, Gen. spec. p. 301.
1854. *C. pygmæa*, Morris, Brit. Foss. p. 7.
1870. *Mantellia pygmæa*, Carruthers, Trans. Linn. Soc. vol. xxvi. p. 703.
Cycadeoidea pygmæa, Schimper, Trait. pal. vég. vol. ii. p. 188.
1893. *C. pygmæa*, Woodward, Lias, p. 378.
1894. *C. pygmæa*, Ward, Proc. Biol. Soc. Washington, p. 80.

The specimen on which Lindley & Hutton founded this species was "communicated by Prof. Buckland, from the Lias at Lyme Regis." The authors note its resemblance to a cone, but they refer to the irregularity of its figure and of the arrangement of the scars, to the presence of a lateral tubercle, and to the absence of seeds as facts indicative of its stem nature. They add, "When cut through from the apex to the base, nothing can be seen except the bases of blunt scales, planted perpendicularly upon a thick and solid centre."

I have not been able to trace the original specimen, and the Museum collection does not include any Liassic stems that can be referred to the same type. So far as one may base an opinion on the drawing and the description, it would appear probable that Lindley & Hutton correctly assigned the specimen to the genus *Cycadeoidea*.

Class CONIFERALES.

Genus **PAGIOPHYLLUM**, Heer.

[Secc. Trab. Geol. Portugal, p. 11, 1881.]

The resemblance between the vegetative shoots usually referred to this provisional and non-committal genus and those of certain recent species of *Araucaria* has led several authors to adopt the name *Araucarites*. It is indeed highly probable that some at least of the fossils placed in the genus *Pagiophyllum* are twigs of Araucarian trees; the abundance of petrified wood exhibiting the characteristic anatomical features of the Araucarieæ lends considerable weight to the inclusion of the Liassic and other species in the genus *Araucarites*. Until we find well-preserved cones in organic union with the twigs, or succeed in ascertaining their anatomical structures, it is, however, safer to retain Heer's generic designation. The coniferous shoots usually assigned to the genus

Pagiophyllum are characterised by their crowded falcate leaves projecting from the branches precisely as in *Araucaria excelsa*. A species from the Lettenkohle of Germany figured by Schütze[1] as *Pagiophyllum Foetterlei*, Stur, hardly conforms to the recognised characteristics of this genus.

Pagiophyllum peregrinum (Lind. & Hutt.).

[Fossil Flora, pl. lxxxviii.]

(Plate V.)

1829. "Plant," De la Beche, Trans. Geol. Soc. vol. ii. [2], p. 29, pl. iv. fig. 7.
1833. *Araucarites peregrinus*, Lindley & Hutton, Foss. Flora, pl. lxxxviii.
1838. *A. peregrinus*, Sternberg, Flor. Vorwelt, vii.–viii. p. 204.
1845. *A. peregrinus*, Unger, Gen. spec. p. 202.
Araucaria peregrina, Kurr, Foss. Flor. Würt. p. 9, pl. i. fig. 1.
1847. *Araucarites peregrinus*, Unger, Chlor. Prot. p. lxxv.
1848. *A. peregrinus*, Bronn, Ind. Pal. p. 91.
1849. *Brachyphyllum peregrinum*, Brongniart, Tableau, p. 104.
Moreauia latifolia, Pomel, Vers. Ges. deutsch. Naturforsch. und Ärzte, p. 21.
1850. *Araucarites peregrinus*, Göppert, Foss. Conif. p. 236.
Cupressus (?) *latifolia*, Buckman, Quart. Journ. Geol. Soc. vol. vi. p. 415, fig. 5.
Araucarites peregrinus, Unger, Gen. spec. p. 382.
1854. *A. peregrinus*, Morris, Cat. Brit. Foss. p. 2.
Cupressus latifolia, ibid. p. 6.
1858. *Araucaria peregrina*, Quenstedt, Der Jura, p. 272, pl. xxxiv. figs. 1–3.
1866. *Cupressus latifolius*, Carruthers, Geol. Mag. vol. iii. p. 545.
Araucaria peregrina, ibid. p. 545.
1867. *Araucarites peregrina*. Moore, Quart. Journ. Geol. Soc. p. 504.
1870. *Pachyphyllum peregrinum*, Schimper, Trait. pal. vég. vol. ii. p. 250.
1876. *P. peregrinum*, Tate & Blake, Yorks. Lias, p. 233.
1884. *P. peregrinum*, Saporta, Plant. Jurass. vol. iii. p. 383, pls. clxxiii.–clxxvi.
1890. *Pagiophyllum peregrinum*, Zittel, Handbuch Palaeont. p. 276, fig. 192.
1892. *Pachyphyllum peregrinum*, Fox-Strangways, Jur. Rocks, vol. i. p. 130.
1893. *P. peregrinum*, Woodward, Lias of England and Wales, p. 378.
1894. *P. liasinum*, Saporta, Flor. foss. Portugal, p. 7, pl. i. fig. 17; pl. ii. fig. 1.
Pagiophyllum peregrinum, Bartholin, Bot. Tidssk. pl. v.
1900. *P. peregrinum*, Ward, U.S. Geol. Surv. p. 308, pl. xlvi.
1903. ? *P. peregrinum*, Möller, Bornholms Foss. Flor. p. 32, pl. v. figs. 14–15.

---

[1] Schütze (01), pl. vi. fig. 1.

Vegetative shoots monopodially branched, bearing crowded, spirally disposed, fleshy leaves. The leaves vary in shape and position; they are usually broadly triangular, sometimes reaching a length of 5 mm., imbricate and fairly closely appressed to the stem; in some shoots they are more open in arrangement and more distinctly falcate. The back of the leaves bears a broad median keel, and the lamina is frequently characterised by numerous longitudinal striations or wrinklings; the apex of the leaf may be obtuse or acuminate.

Flowers not known.

Among the specimens included under this type one notices differences as regards the shape and disposition of the leaves which, however, are not of sufficient importance to justify the use of more than one specific name.

The occurrence of leaves having different shapes on branches of the same tree is not an uncommon phenomenon among recent Conifers, and there can be little doubt that too much weight has been attached to slight variations as regards leaf-form in the determination of fossil coniferous twigs. *Araucaria*, perhaps, is more constant in the shape and disposition of its leaves than some other recent genera, but cases of heterophylly[1] are by no means unknown.

It is probable that *Pagiophyllum peregrinum* represents the vegetative branches of an Araucarian tree. Lindley & Hutton recognised the resemblance of the Lias specimens from Lyme Regis to *Araucaria excelsa*, and adopted the generic name *Araucaria*; Unger, Kurr, Göppert, and some other authors have also used this generic name or *Araucarites*. A good drawing by Adolphe Brongniart of an English specimen from Lyme is published by Saporta in the third volume of his *Plantes Jurassiques*; Saporta notices a difference as regards habit between the fossil branches and those of recent Araucarias. It is true that the habit of branching, as shown in Pl. V. Fig. 4, and in the specimen figured by Saporta, is not quite the same as in *Araucaria excelsa*, but the recent species *A. brasiliensis* agrees closely with the Liassic form as regards the disposition of the smaller branches.

---

[1] Siebold (70), pl. 140; Potonié (99), p. 292, fig. 295.

A specimen from Lyme Regis in the Jermyn Street Museum bearing the name *Cupressites liasinus*, Quenst., is probably a fragment of *Pagiophyllum peregrinum*.

It is hardly possible to distinguish between *P. peregrinum* and various species[1] described from higher horizons in the Jurassic system. There can be little doubt that the *Araucaria*-like branches indicate the abundance, during the greater part of the Jurassic era, of Coniferous trees in all probability nearly related to the modern Araucarias.

38,349. Pl. V. Fig. 1.

This specimen illustrates a broader form of shoot with large fleshy and angular leaves, which have a somewhat more open and spreading arrangement than in most of the examples of this species.

Lyme Regis. *Bean Coll.*

52,665*a*. Pl. V. Fig. 2.

Part of a branched specimen showing the falcate form and longitudinal ribbing of the leaves.

Lyme Regis. *Purchased*, 1870.

35,044. Pl. V. Fig. 3. (Four times nat. size.)

Portions of two leaves of a twig 6·5 cm. long, similar to the specimen shown in Fig. 2. The fleshy leaves are well preserved, showing a median keel and numerous small papillæ, which occasionally occur in regular longitudinal rows.

Lyme Regis. *Purchased*, 1876.

40,675. Pl. V. Figs. 4, 4*a*.

A larger specimen, 19 cm. long, illustrating the branching habit. The enlarged leaf (Fig. 4*a*) shows the median keel and irregular longitudinal ribbing.

Lyme Regis. *Purchased*, 1859.

V. 9608. Pl. V. Fig. 5. (Four times nat. size.)

An unbranched fragment 8 cm. long, bearing strongly wrinkled leaves.

Lyme Regis. *Purchased*, 1855.

---

[1] Cf. Saporta (84), pls. 178–183.

52,665. Pl. V. Fig. 6. (Five times nat. size.)

A piece of a branched shoot, bearing crowded imbricate leaves, and differing slightly in appearance from the form represented in Fig. 1; similar to **40,675** (Pl. V. Fig. 4).

Lyme Regis. *Purchased*, 1870.

45. Figured by De la Beche (Trans. Geol. Soc. vol. ii. [2], pl. iv. fig. 7). A well-preserved fragment 6 cm. long and 8 mm. broad; the leaves have broadly triangular apices and are slightly falcate.

Lyme Regis. *Presented by Sir Henry T. De la Beche*, 1837.

36,655. A branch with more spreading leaves than in De la Beche's specimen, and closely resembling the recent *Araucaria excelsa*, R. Br.

Lyme Regis. *Purchased*, 1870.

52,665*b*. The large broad leaves, with a longitudinally wrinkled lamina, are well preserved.

Lyme Regis. *Purchased*, 1870.

52,680. Large branched specimens, illustrating on the same shoot the compact form like that represented in Pl. V. Fig. 4, and in De la Beche's figure, in addition to the more spreading and open arrangement of the leaves shown in Fig. 1, Pl. V.

Lyme Regis. *Purchased*, 1865.

52,693. An unusually thick piece, 18 cm. long and 1·3 cm. broad; leaves of the longer and more tapering form.

Lyme Regis. *Purchased*, 1873.

Other specimens:—**V. 41, 35,045, 36,656, 36,657, 36,658, 36,659, 40,676, 40,677** (good branched specimen), **40,679, 52,857, 52,859.**

V. 41 *presented by Capt. H. G. Lyons*, 1882; *the others purchased.*

PLANTA INCERTÆ SEDIS.

[*Strobilites elongata*, L. & H. (Foss. Flor. pl. lxxxix. 1833).]

1833. *Strobilites elongata*, Lindley & Hutton, Foss. Flor. pl. lxxxix.
1848. *S. elongata*, Bronn, Ind. Pal. p. 1203.
1849. *Pinites* (?) *elongatus*, Brongniart, Tableau, p. 104.
1850. *P. elongatus*, Unger, Gen. spec. p. 361.
P. elongatus, Göppert, Foss. Conif. p. 223.
1854. *Strobilites elongata*, Morris, Brit. Foss. p. 23.
1866. *P. elongatus*, Carruthers, Geol. Mag. vol. iii. p. 544.

The specimen from the Blue Lias of Lyme figured under the above name by Lindley & Hutton presents the appearance of an imperfect and partially destroyed, long and narrow cone with large imbricate scales. In the original description it is suggested that the cone may have been borne by *Pagiophyllum peregrinum* (L. & H.), but, as Carruthers[1] says, it is impossible to determine the nature of the fossil from the published drawing.

CONIFEROUS WOOD.

a. Casts.

The occurrence of driftwood in the Liassic rocks of the Dorsetshire coast has been mentioned by De la Beche[2] and other authors.[3] The specimens occur in the form of casts, many of which present the appearance of petrified wood, but unfortunately the internal structure has not been preserved, and microscopical examination is not possible. Several pieces of branches have been found which reach a length of three or four feet; with the casts are often associated the shells of Ammonites, and these occasionally occur attached to the surface of the fossil wood.[4] The branches may belong to the tree of which the smaller twigs are known as *Pagiophyllum peregrinum*, but we lack the necessary data for the determination of the leafless casts.

V. 334. Text-fig. 4. A branched specimen 4 feet 6 inches in length in association with *Ægoceras planicostatum*, Sow.

Lyme Regis. *History unknown.*

Other specimens from Lyme Regis and elsewhere:—**V. 171**, small piece of wood from Whitby, Yorkshire. **V. 1073**, a large branched specimen, 45 cm. in length, from Lyme Regis. **V. 4078**, two large pieces, one of which bears several branch-scars; Lyme. **V. 4079**, **V. 4080**, Lyme. **10,354**, fragment from Charmouth, Dorsetshire (*Mantell Coll.*). **11,235**, fragment from Litchborough, Northamptonshire (*Mantell Coll.*). **11,255**, Charmouth (*Mantell Coll.*). **18,403**,

---

1 Carruthers (66).

2 De la Beche (24).

3 Parkinson (11), p. 375; Woodward (93), p. 8.

4 For a figure of a piece of Liassic driftwood (specimen No. 52,894) bearing a shell of *Ægoceras planicostatum*, see Seward (98), p. 61, fig. 7.

20,460, Crick, Northamptonshire. **20,461**, Kilsby tunnel, Northamptonshire. **20,464**, Warkworth, Northamptonshire.

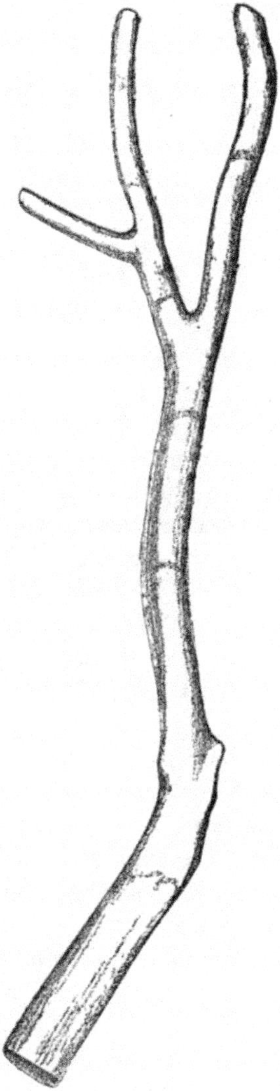

FIG. 4.—Cast of Coniferous stem from Lyme Regis. V. 334. ⅛ nat. size.

20,466, Bugbrook, Northamptonshire (age doubtful). **20,468,** Crick. **20,469,** Kilsby tunnel. **20,473,** Staverton, Northamptonshire. **20,475,** Kilsby tunnel. **20,482,** Litchborough. **20,483,** Roade, Northamptonshire. **20,486,** Northampton (partly converted into jet). **20,488,** Kilsby tunnel. **20,492,** Bugbrook. **20,909,** Braunston, Northamptonshire (this and preceding thirteen specimens from *Miss Baker's Coll.*). **40,671,** Whitby (partly converted into jet), (*Bean Coll.*). **52,623,** Whitby. **52,625,** Whitby. **52,671,** near Grantham. **52,842, 52,857,** Charmouth, Dorsetshire. **52,894,** Lyme Regis. **52,902,** eight feet long; Lyme Regis.

b. Petrifactions.

(Plates VI.–VII.)

Although a considerable amount of driftwood occurs in the Lower Lias beds near Lyme Regis, it is practically valueless botanically, as the tissues have not been petrified. From the Lias of Yorkshire several specimens have been obtained in which the tissues are exceedingly well preserved. Tate & Blake[1] mention the occurrence of fossil wood at various horizons in the Yorkshire Lias, and Fox-Strangways[2] speaks of wood in the zone of *Ammonites capricornus*, *A. margaritatus*, and elsewhere.

In his classic work on *The Internal Structure of Fossil Vegetables* Witham[3] describes and figures sections of silicified wood from Upper Liassic beds near Whitby, from material supplied by Mr. Nicol. He recognises two species which he names *Peuce Lindleiana* and *Peuce Huttoniana.* The wood referred to the former species is undoubtedly Araucarian; possibly *P. Huttoniana* may also belong to the same type, but as it is not clear from Witham's account in what respect the two forms differ from one another, I propose to retain only the specific designation *Lindleiana,* or rather *Lindleii.* Both names occur in Witham's book, but the latter is selected as being more in accordance with our present system of nomenclature.

---

1 Tate & Blake (76), pp. 39, 64, 153.

2 Fox-Strangways (92), *passim.*

3 Witham (33).

In addition to these two forms of *Peuce*, Witham gives drawings of other specimens of coniferous wood from the Whitby Lias to which he does not apply specific names. In an important paper published in 1834, Nicol[1] notes the occurrence of three types of coniferous wood in the Whitby Lias, and recognises certain histological characters which demonstrate the Araucarian nature of such specimens. Nicol, like many subsequent writers, states that the wood of existing species of *Araucaria* is peculiar in the absence of annual rings. This statement, though partially true, is by no means entirely correct, as distinct annual rings do occur in some modern Araucarias. The same author correctly describes the contiguous arrangement and polygonal form of the bordered pits as an Araucarian characteristic.

My examination of numerous sections of wood from the neighbourhood of Whitby and Staithes in the Museum collection has enabled me to recognise two well-marked types, one of which is an undoubted Araucarian type, while the other may be included in Göppert's comprehensive genus *Cupressinoxylon*. It is probable that more than two species are represented, but the material available is hardly sufficient to warrant the institution of additional specific terms.

Genus **ARAUCARIOXYLON**, Kraus.

[Schimper, Trait. pal. vég. vol. ii. p. 380, 1872.]

This generic designation was formerly made use of in the descriptions of wood from Palæozoic as well as from Mesozoic and Tertiary rocks; but as the Palæozoic specimens are in many cases known to belong to *Cordaites* and other genera which differ in important features from true Conifers, it is better to restrict the name *Araucarioxylon* to fossil wood which we believe to belong to trees closely allied to existing members of the Araucarieæ. In thus limiting the use of *Araucarioxylon* I am following the example of Knowlton,[2] Zeiller,[3] Scott,[4] and other authors.[5]

---

[1] Nicol (34).
[2] Knowlton (89), (90).
[3] Zeiller (00), p. 280.
[4] Scott (99), (02).
[5] Penhallow (00).

Araucarioxylon Lindleii (Witham).

[Internal Struct. Foss. Veg. p. 58, pl. ix. figs. 1–5; pl. xv. figs. 1–3.]

(Plate VI.; Pl. VII. Figs. 2, 3, 5.)

1833. *Peuce Lindleiana*, Witham, Int. Struct. Foss. Veg. p. 58, pl. ix. figs. 1–5; pl. xv. figs. 1–3.
1845. *P. Lindleyana*, Unger, Synopsis, p. 207.
1847. *P. Lindleyana*, Unger, Chlor. protog. p. 35.
1848. *P. Lindleyana*, Bronn, Ind. Pal. p. 951.
1849. *P. Lindleyana*, Brongniart, Tableau, p. 106.
1850. *P. Lindleyana*, Unger, Gen. spec. foss. p. 370.
Pinites Lindleyanus, Göppert, Foss. Conif. p. 217.
1854. *Peuce Lindleyana*, Morris, Brit. Foss. p. 17.
1866. *Pinites Lindleyanus*, Carruthers, Geol. Mag. vol. iii. p. 545.
1876. *Peuce Lindleyana*, Tate & Blake, Lias, pp. 233, 474.
1882. *P. Lindleyana*, Etheridge, Quart. Journ. Geol. Soc. vol. xxxviii., Proc. p. 140.

Annual rings distinct. Wood homogeneous, composed of tracheids bearing 1–3 rows of bordered pits on their radial walls; the pits usually occur in two contiguous rows, alternately disposed and polygonal in shape; occasionally three rows of pits occur, and some of the tracheids possess a single row (Pl. VII. Fig. 5). Medullary rays distinct in transverse section of the wood, composed of narrow, radially elongated cells, with single pits in their walls; the rays vary in depth from a single row to more than twelve rows of parenchymatous cells. In a tangential section of the wood the rays occur as long and narrow groups one cell in breadth; occasionally a ray is seen to be two cells in breadth. Pith composed of comparatively large polygonal cells, passing into a fairly distinct perimedullary zone of smaller and darker cells (Pl. VI. Fig. 3).

51,484. Pl. VI. Figs. 1, 3, and 4; Pl. VII. Fig. 3.

A large transverse section, 8·5 × 5·5 cm., cut from the same piece of wood as that which afforded the section figured by Witham in his pl. ix. fig. 1. The pith, which is imperfectly preserved and occupies an excentric position, consists mainly of fairly large polygonal cells (Pl. VI. Fig. 3), passing into a perimedullary zone of smaller cells. Groups of darker elements in the pith region may represent sclerous nests, but the preservation is not sufficiently good to enable one to distinguish between dark cell-contents and thickened membranes.

The wood abuts on the perimedullary zone in the form of bluntly terminated wedges. The annual rings are very distinct (Pl. VI. Fig. 1); those on one side of the pith vary in width from 1 to 2 mm., while on the other side they reach a breadth of 3·5 mm.: this feature bears an interesting resemblance to the inequalities frequently seen in the wood of recent trees which are due to the more vigorous development of tissue on that side of the stem exposed to the more favourable quarter. In one part of the section there appears to be an indication of a double annual ring, represented by a narrow zone of smaller tracheids in the spring wood. The medullary rays are clearly seen in Pl. VI. Fig. 1 as dark bands, and in Fig. 5, Pl. VII. two rays are shown in surface-view; they consist of radially elongated parenchymatous cells which occasionally exhibit small simple pits, and the cavities are often full of vacuolated brown contents. The rays vary in depth from one to fourteen rows of cells.

The wood appears to be without resin-canals, but in one or two places there occurs a series of irregular cavities—as shown at *r*, Fig. 3, Pl. VII., and in longitudinal section in Fig. 2, *c*, Pl. VI.—bounded by dark-brown parenchymatous cells with convex walls bulging into the oval or circular spaces. These spaces are not regular enough to be compared with the resin-canals of the Abietineæ, and their occurrence is exceptional and local. In the portion of the stem represented in Fig. 3, Pl. VII., they occur in the spring wood; in this interrupted series of spaces one is able to trace the development of the irregular cavities from patches of dark-brown cells, which gradually separate to form a space into which resin may have been secreted. Seen in longitudinal section (**51,724**, Pl. VI. Fig. 2), the cavities are bounded by one or two rows of short parenchymatous cells, and these short elements are in places separated from the tracheids by a row of long and narrow cells. In all probability these spaces may be regarded as the result of abnormal conditions, which have caused the formation of a resinous secretion. A similar occurrence of abnormal secretory parenchyma is figured by Conwentz[1] in the wood of *Pinites* preserved in Baltic amber.

Whitby, Upper Lias. *Bryson Coll.*

---

[1] Conwentz (90), pls. vi.–viii.

51,724. Pl. VI. Fig. 2.

A radial longitudinal section showing medullary rays and the patch of secretory tissue (*r*) seen in the transverse section reproduced in Fig. 3, Pl. VII. The radial walls of the tracheids bear contiguous rows of bordered pits, as is seen more distinctly in Fig. 5.

Whitby, Upper Lias. *History unknown.*

51,488. Pl. VII. Fig. 5.

Radial longitudinal section showing very clearly the characteristic Araucarian form and disposition of the bordered pits. In some of the medullary rays seen in this section (not shown in the photograph) there occur small and simple pits. In one or two places long and narrow parenchymatous cells occur at the limit of the autumn wood. *Bryson Coll.*

51,484. Pl. VI. Fig. 4.

Radial longitudinal section showing the shape and vacuolated contents of the medullary-ray cells. *Bryson Coll.*

Other sections :—**51,437.** A large transverse section 6·5 × 5 cm.; the pith, composed of large polygonal cells, is partially preserved, as in **51,484** (Pl. VI. Fig. 3). Annual rings well defined and wide, reaching a breadth of 8 mm. Medullary rays distinct.

Scarborough. *Bryson Coll.*

51,450. No doubt cut from the same specimen as **51,486.** Radial longitudinal section; preservation poor, but contiguous rows of polygonal pits are seen here and there.

Whitby, Upper Lias. *Bryson Coll.*

51,452. Radial longitudinal section showing clearly marked annual rings. Medullary rays vary in depth from one to twelve or more rows. A row of long and narrow cells occurs in one place next the spring tracheids at the limit of the autumn wood.

Whitby. *Bryson Coll.*

51,473. Transverse section. In all probability specifically identical with the preceding sections.

Staithes, Upper Lias. *Bryson Coll.*

51,486. Tangential longitudinal section. Medullary rays usually one cell broad, but occasionally two cells wide; the number of cells in a ray varies from one to twenty. Probably a section of *Araucarioxylon Lindleii.*

Whitby, Upper Lias. *Bryson Coll.*

51,489. Transverse section.

Whitby, Upper Lias. *Bryson Coll.*

51,490. Radial longitudinal section; preservation very good. The pits are particularly well shown, usually occurring in two or three contiguous rows, but occasionally in four rows. Pits shown on the walls of some of the medullary-ray cells.

Whitby. *Bryson Coll.*

51,697. Tangential longitudinal section, probably from the same specimen as **51,486.**

51,702. Badly preserved transverse section, probably *Araucarioxylon Lindleii.*

Upper Lias. *Bryson Coll.*

Araucarioxylon Lindleii (?).

51,449. Pl. VII. Fig. 2.

Transverse section characterised by the great breadth, if not by the absence, of annual rings. The tissue is crushed in places, hence the difficulty of stating definitely that rings of growth do not occur. Medullary rays comparatively broad and well marked. This wood bears a close resemblance to that which is associated with some of the specimens of jet described below.

Whitby, Upper Lias. *Bryson Coll.*

51,487. A transverse section, no doubt identical with **51,449** (Pl. VII. Fig. 2). No rings visible.

Whitby. *Bryson Coll.*

51,491. Oblique tangential section, showing traces of contiguous and polygonal pits on some of the tracheids.

Whitby. *Bryson Coll.*

51,494. A crushed transverse section.

Near Whitby. *Bryson Coll.*

51,647. Badly preserved transverse section.

Staithes, Upper Lias. *Bryson Coll.*

51,728. Radial longitudinal section.

Whitby. *Bryson Coll.*

Genus CUPRESSINOXYLON, Göppert.

[Foss. Conif. p. 196, 1850.]

As Knowlton points out, this generic name has been used in a comprehensive sense; he remarks—"As our knowledge of the internal organization of living Conifers has been gradually worked out by Göppert, Kraus, Beust, and others, the fact has become more and more apparent that types founded upon these characters alone must be regarded in a measure as comprehensive." [1]

Schenk defines *Cupressinoxylon* as follows:—"Annual rings distinct, narrow. Resiniferous parenchyma abundant. Cells of the medullary rays thin-walled, with simple pits in their radial walls. Types included under this head: Cupressaceæ, Podocarpeæ, *Cunninghamia*, Taxodineæ, *Phyllocladus*, *Dacrydium*, *Ginkgo*, *Saxegothea*, *Abies Webbiana*." [2]

In 1898 Mr. C. A. Barber [3] published a careful description of a type of Lower Greensand fossil wood from the Isle of Wight under the name *Cuppressinoxylon vectense*, and discussed the value of the histological characters on which this and other genera have been founded.

The material which I have described from the Lias of Whitby is too meagre to admit of a complete diagnosis, and it is with some hesitation that the generic name *Cupressinoxylon* has been selected. One characteristic feature of this genus is the presence of resin-cells among the tracheids, but I have not recognised in the few available sections any certain indications of resiniferous parenchyma.

Cupressinoxylon Barberi, sp. nov.

(Pl. VII. Figs. 1, 4, and 6.)

The sections referred to this species exhibit certain features which distinguish the wood from that of *Araucarioxylon*; for the sake of convenience I have designated this type *C. Barberi*, after my friend Mr. C. A. Barber, the author of a thorough and critical

---

[1] Knowlton (89), p. 43.
[2] Schenk, in Zittel (90), p. 862.
[3] Barber (98).

paper on a species of *Cupressinoxylon* from the Lower Greensand of the Isle of Wight. The chief characters may be summarised as follows:—Annual rings very distinct. Tracheids wider than those of *Araucarioxylon Lindleii*. Medullary rays composed of long and narrow cells, consisting occasionally of more than twenty cell-rows as seen in surface-view. Bordered pits occur either in a single row (in which case they may be in contact or separate and irregularly arranged) or frequently in two rows; the pits of the adjacent rows are either alternate or on the same level. No resin-canals or resiniferous tissue.

51,498. Pl. VII. Fig. 1.

Radial longitudinal section showing numerous narrow rings of growth. This slice was no doubt cut from the same specimen as 51,495 and 51,496. The bordered pits often occur in a single row, either contiguous or separate, or frequently in two rows, the individual pits being placed alternately or opposite to one another. The pits are circular or oval, and not polygonal, and do not appear to exceed two rows on any one tracheid.

Whitby, Upper Lias. *Bryson Coll.*

51,496. Pl. VII. Fig. 4.

A transverse section 5 cm. in width, reproduced slightly less than natural size. Annual rings (Pl. VII. Fig. 6) clearly marked and narrow, more abrupt than in *Araucarioxylon Lindleii*, and characterised by a less gradual decrease in the diameter of the spring tracheids as they pass into the narrow elements at the limit of the rings. Medullary rays narrow and less distinct than in the preceding species.

Whitby, Upper Lias. *Bryson Coll.*

51,495. Pl. VII. Fig. 6.

Similar to the section shown in Fig. 4; the autumn wood is composed of a few very narrow tracheids, which are succeeded rather suddenly by much larger elements.

Whitby, Upper Lias. *Bryson Coll.*

Other specimens:—**V. 328.** Transverse section; annual rings well defined, from 1 to 2 mm. broad. Structure precisely the same as in 51,495 and 51,496.

Whitby.

51,447. Longitudinal radial section; several deep medullary rays.

51,497. Large and badly preserved transverse section.

51,757. A transverse section of inferior preservation; possibly referable to *Cupressinoxylon Barberi*.

Near Whitby, Upper Lias. *Bryson Coll.*

CONIFEROUS WOOD OF DOUBTFUL POSITION.

51,500. Tangential section, including a parenchymatous region in the neighbourhood of a branch.

Whitby. *Bryson Coll.*

51,694. Transverse section, showing exceptionally narrow rings of growth; much crushed.

Staithes. *Bryson Coll.*

51,758. Badly preserved radial section, labelled "Fossil tree, Whitby, Tertiary." Probably a piece of Liassic wood.

Bryson Coll.

JET.

(Plate VIII.; Text-figs. 5–7.)

The name 'jet' is said to be derived from the river Gagas in Lycia, where the substance was obtained. The following passage from Pliny is of interest in reference to the origin of the term jet and as regards its supposed properties:—

"Gagates lapis nomen habet loci et amnis Gagis Lyciæ. Aiunt et in Leucolla expelli mari atque intra xii stadia colligi. Niger est, planus, pumicosus, levis, non multum a ligno differens, fragilis, odore, si teratur, gravis. Fictilia ex eo scripta non delentur; cum uritur, odorem sulphureum reddit; mirumque, accenditur aqua, oleo restinguitur. Fugat serpentes ita recreatque volvæ strangulationes. Deprendit sonticum morbum et virginitatem suffitus. Idem ex vino decoctus dentibus medetur strimisque ceræ permixtus. Hoc dicuntur uti Magi in ea, quam vocant axinomantiam, et peruri negant, si eventurum sit quod aliquis optet."[1]

---

[1] Pliny, Nat. Hist. Book xxxvi. 34, 141.

The occurrence of ornaments in barrows of the Bronze Age affords evidence of the great antiquity of jet as a material used for decorative purposes.

Mr. Fox-Strangways has given an interesting account of the references to jet in ancient literature: he quotes from the rolls of Whitby Abbey for 1394, and the following lines from Cædmon, who died about 670 or 680, and was buried in Whitby Abbey:—

"Jeat stone, almost a gemm, the Lybians find,
But fruitful Britain sends us wondrous kind;
'T is black and shiny, smooth and ever light;
'T will draw up straws, if rubbed till hot and bright;
Oyl makes it cold, but water gives it heat."[1]

Standing on the cliffs above Whitby, on the Yorkshire coast, one frequently notices a line of breakers parallel to the coast which mark the position of a ledge of hard rock a short distance from the coastline. This is the jet-rock of the Upper Lias series from which the Whitby jet was obtained. The jet-rock crops out at various localities in the neighbourhood of Whitby, and jet has been worked in Eskdale, Darbydale, and in several of the other dales that intersect the moorlands of East Yorkshire.[2] In *The Organic Remains of a Former World*, published in 1811, Parkinson speaks of jet as *Succinum nigrum*, a term used by several ancient writers, and expresses the opinion that in some cases jet was originally wood, as shown by its fibrous texture.[3] In a work by Young, entitled *A History of Whitby and Streoneshalk Abbey*, jet is described as a substance which "may be properly classed with fossil wood, as it appears like wood in a high state of bituminization."[4] Jet occurs in masses varying from ½ to 1 inch in thickness and from 3 to 18 inches broad, occasionally reaching a length of 12 feet. Young also mentions the occurrence of specimens of jet containing a core of silicified wood. Similar specimens are described by Young & Bird, in their volume on the Yorkshire coast,[5] in which the jet itself forms

---

1 Fox-Strangways (92), p. 455.
2 Fox-Strangways, Reid, & Barrow (85), p. 18.
3 Parkinson (11), p. 232.
4 Young (17), p. 783.
5 Young & Bird (22), p. 188.

a crust on the outside of a log of pyritised or silicified wood. Witham speaks of jet as possessing a "ligneous structure," and as exhibiting "concentric rings of fibrous tissue." He was the first to give a figure of a section of jet showing indications of a woody structure.[1]

An interesting paper on Whitby jet and its manufacture, by Mr. J. A. Bower, was published in the *Journal of the Society of Arts* in 1873.[2] This author deals at length with the history of the working of jet, and briefly discusses its probable origin. He notes that most of the jet-workers regard it as having a ligneous origin. Bower does not consider that vegetable matter is essential for its production, and in support of this opinion he mentions the occurrence of bones and fish-scales which have been converted into jet. Jet, he says, has the appearance of a substance distilled from the rock, which has in some cases impregnated vegetable matter and animal substances, or has simply consolidated in fissures. On the whole he considers that the evidence supports the view that jet has been formed as a distillate from the jet-rock.

Messrs. Tate & Blake have dealt with the origin of jet in their book on the Yorkshire Lias: they point out that the beds in which it occurs are highly charged with bitumen, and they consider that the jet owes its origin to "the segregation of the bitumen in the intervals of the shales, which, allowing to a certain extent the access of air, has hardened into jet, a process which may undoubtedly be now going on." They go on to say, "There seems to be no reason whatever for connecting it with wood, beyond its having a remotely similar composition, though, of course, we have thrown no light on the cause of the presence of the bitumen itself."[3]

Jet is described by Messrs. Fox-Strangways & Barrow, in the Geological Survey Memoir on the Coast between Whitby and Scarborough, as water-logged Coniferous wood "from which in a few rare cases all trace of structure has been removed."[4] In a more recent work, Fox-Strangways points out that it is very certain that jet is derived from vegetable matter in some form or

---

[1] Witham (33), pp. 10, 50, pl. xi. fig. 3.
[2] Bower (73).
[3] Tate & Blake (76), p. 179.
[4] Fox-Strangways & Barrow (82), p. 21.

other. He considers that it was not formed from plants which grew where the jet is found, as it does not occur in regular seams, but in isolated blocks. It is more probable, he adds, that jet was produced from trees which had been floated into their present position, buried in Lias mud, and subsequently converted into bitumen, "which became diffused through the neighbouring shale or occupied cavities in them, where it is now found as jet, often occurring as pseudomorphs of organic remains." [1]

This brief historical review suffices to show that the balance of opinion is in favour of a vegetable origin for jet. Several writers, however, prefer to regard the masses of flattened jet, having the form and texture of wood, as pieces of stems or branches impregnated with a fluid bitumen, rather than as pieces of wood which have been directly converted into jet. With the exception of Witham, none of the authors who have discussed the origin of jet appear to have made a microscopical investigation of its structure.

The result of an examination of several sections of Whitby jet in the British Museum collection leads me to express the opinion [2] that jet has been formed by the alteration of wood: the masses of wood with the form and structure of jet that were formerly extensively worked as the source of that material, probably represent portions of trees the tissues of which underwent certain chemical changes resulting in the partial or complete obliteration of the vegetable structure and in the substitution of jet for wood.

In all probability Whitby jet has been produced in large measure by the alteration of wood of the Araucarian type. Undoubted *Araucarioxylon* wood occurs in the Lias rocks in association with the jet (Plate VI.), and the structure of the tissues met with in specimens like those shown in Pl. VIII. Figs. 1, 2, 4, and 6, appears to be of the same type. It is interesting to find that Knowlton speaks of jet found in the Richmond Basin, Virginia, as having been formed from *Araucarioxylon* wood.[3]

A section cut from a good sample of hard Whitby jet is of a rich sherry or golden yellow colour. Microscopical examination may

---

[1] Fox-Strangways (92), p. 455.

[2] Briefly stated in a note published in 1901; see Seward (01), p. 856.

[3] Knowlton, in Shaler & Woodworth (97), p. 517.

reveal no definite structure in the homogeneous substance, or, more frequently, the matrix is seen to be traversed by a number of dark-brown undulating, zigzag, or roughly parallel lines (Pl. VIII. Figs. 5, 7); while in other sections the matrix exhibits a series of faint vertical lines crossed here and there at right angles by short dark lines (Pl. VIII. Fig. 3). A transverse section of jet usually appears as represented in Figs. 5 and 7, Pl. VIII.; the contorted dark lines reveal no indication of their origin or meaning, and the rich yellow or orange matrix shows nothing more than an indistinct spotted appearance too faint to afford a clue as to its origin.

It is interesting to find that a microphotograph of a section such as I have described reveals more distinct evidence of structure in the matrix, and brings to light more definite though indistinct traces of a fine net-like texture suggestive of compressed plant-tissue, than the section itself. A section cut parallel to the edge of a compressed mass of jet—that is, a tangential section—exhibits a number of slightly irregular vertical lines crossed at intervals by series of short dark patches (Pl. VIII. Fig. 3). This structure is obviously a faint representation of that of a coniferous stem, the vertical lines being the walls of the tracheids, while the darker patches mark the position of the medullary-ray cells. Fig. 4, Plate VIII. represents a specimen one edge of which is composed of pure jet having the structure seen in Figs. 5 and 7; the rest of the slice consists in part of mineral material, showing little or no trace of woody structure, with occasional patches of fairly well preserved petrified wood cut across transversely.

Figs. 1 and 2 represent portions of sections which are in part composed of jet and in part of silicified coniferous wood; these preparations show the extension of the dark and apparently structureless jet into the tissues of the wood, and in the wood itself one sees that the dark lines follow the course of the radial walls of the tracheids. Similarly, in Fig. 6 we have a piece of coniferous wood in transverse section invaded by dark lines which clearly follow the course of the rows of tracheids and extend along their radial walls. Text-fig. 6 shows a small piece of wood in transverse section of which the tracheids are more or less ragged and disorganised, the tangential walls have partially broken down, while the radial walls are being replaced by a brown substance which forms conspicuous dark bands (*a*). Identical dark bands are seen in Text-fig. 7 formed along the radial walls of the tracheids; the rest of the section presents the same spotted appearance as that which characterises the yellow matrix of pure jet. The cavities of the tracheids become filled with an orange-coloured substance, and the outlines of their walls may be faintly traced in a microphotograph of a section of ordinary jet.

The contorted appearance of the dark lines shown in a transverse section of jet points to contraction of the wood during its conversion into jet. I have no hesitation, then, in describing the dark lines seen in Figs. 5 and 7, Pl. VIII. as having been formed by the production of a deep brown substance along the course of the radial walls of the tracheids, their wavy or zigzag form being due to

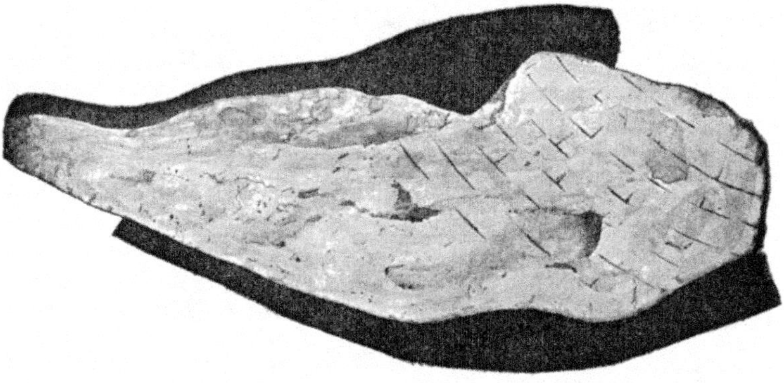

FIG. 5.—Jet partially enclosing a core of stone consisting of an imperfectly preserved cast of a piece of wood. (Whitby Museum.)

contraction. The mottled appearance of the lighter-coloured matrix is due to the persistence of faint traces or 'ghosts' of the membranes of wood-elements. The flattened form of the large blocks of jet met with in the jet-rock is no doubt very largely due to the compression of the wood in the process of its alteration into jet. It has been pointed out by more than one writer that specimens are not uncommon in which a thin layer of jet forms a crust over a core of stone. In specimens of this kind the thickness is considerably greater than in pieces of ordinary jet.

Text-fig. 5 represents a diagrammatic sketch of an end-view of a specimen in the Whitby Museum, which consists for the most part of a central core of sandstone showing in places distinct indications of concentric rings traversed here and there by crystalline lines which mark the position of medullary rays. In this case the outer portion of the block of wood has been converted into jet, while the inner portion decayed and was replaced by some sedimentary material which has partially retained the form of the original wood-structure. The replacement by the sedimentary material was not accompanied by the compression which characterised the conversion of the wood into jet, hence the greater breadth of specimens such as that shown in Text-fig. 5, as compared with those which consist wholly of jet.

The Whitby Museum contains some specimens of jet in which Belemnites are partially embedded, clearly pointing to a soft condition of the substance in which the pointed end of a belemnite became firmly enclosed. There is no reason for supposing that in this case the jet was not formed from wood; it may indeed have been produced from animal matter, but on the other hand it is not improbable that the altered woody tissue may have been in a soft condition and in a state in which a hard substance may easily have sunk into it.

DESCRIPTION OF SPECIMENS.

(All the examples are from the Upper Lias of Whitby, and from the *Bryson Coll.*)

51,601. Pl. VIII. Fig. 1.

This section has been cut from a specimen consisting in part of normal jet of a deep sherry colour and exhibiting the characteristic structure like that shown in Figs. 5 and 7. A portion of the specimen consists of fairly well preserved silicified coniferous wood. The importance of this section is the transition shown between the wood and the jet; at the junction between the two, as seen in the photograph, the conversion of the wood into jet is seen to take place first between the radially disposed tracheids. Cf. Text-figs. 6 and 7 and Fig. 2, Pl. VIII. This section and that numbered 51,451 were probably cut from the same specimen.

51,639. Pl. VIII. Fig. 2.

A piece of silicified wood in transverse section showing very faint annual rings and suggesting an Araucarian species. Like the slice represented in Fig. 1, this also shows the gradual encroachment of the jet along the rows of tracheids. In sections such as those shown in Figs. 1 and 2 it is not difficult to trace the dark lines and patches, clearly shown in the photographs as occurring in the wood, into the wavy and contorted bands which are seen in the portions of the specimens already converted into jet.

51,595. Pl. VIII. Fig. 3.

An obliquely tangential section of pure jet of a rich sherry colour. The medullary rays are shown as superposed rows of dark patches and in some places as more radially extended lines. Both tracheids and medullary-ray cells are represented by faint lines or darker patches, and afford a blurred and obscure reproduction of the actual wood.

51,493. Pl. VIII. Fig. 4.

The section, slightly reduced in the photograph, is 4·5 cm. in length: it consists of bands and patches of ordinary jet interspersed with badly preserved silicified wood seen in transverse section. The jet presents the same appearance as that represented in Figs. 5 and 7. In many parts of the section the passage from wood to jet is clearly shown.

51,623. Pl. VIII. Figs. 5 and 7.

Sections cut from a piece of pure jet, which has a rich brown colour in transmitted light. The lighter-coloured matrix is traversed by dark opaque bands exhibiting small zigzag contortions (Fig. 7) and large folds. The matrix between the bands resolves itself, under a high magnifying power, into a network of more or less rectangular spaces representing tracheids, the cavities of which are bounded by very fine lines. Under a lower power the matrix appears to be without structure and homogeneous. The wood from which the jet has been formed is cut across transversely; the dark bands probably mark the position of the medullary rays, and their junction can be readily traced in such specimens as those represented in Figs. 1 and 2; the mottled appearance of the matrix, as shown in the photographs (Figs. 5 and 7), is simply the faint remnant of the wood.

51,448. Pl. VIII. Fig. 6.

A small piece of a specimen similar to that represented in Fig. 4, consisting of a dark patch of jet and portions of silicified wood in transverse section. The figured piece shows the beginning of the formation of the dark bands (*a*) between the tracheids—possibly occupying the position of the medullary rays. The same specimen shows also that the dark bands in the jet are parallel to the rows of tracheids (this is not seen in the photograph, but is readily recognised on a careful examination of the actual section). The Text-figs. 6 and 7 were drawn from this section.

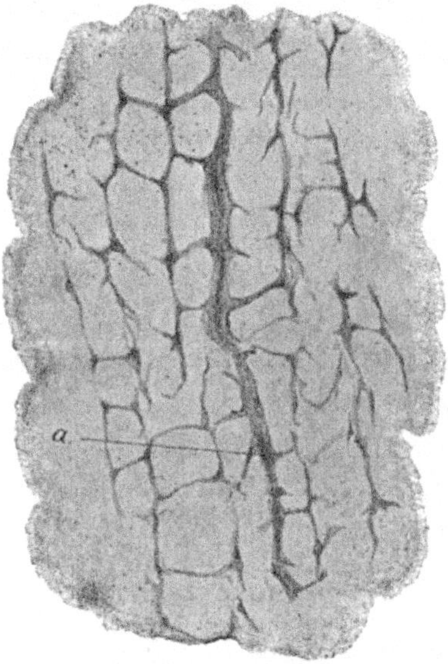

FIG. 6.—Transverse section of coniferous wood passing into jet. Highly magnified. 51,448.

Other specimens:—**51,451.** This section and 51,639 were probably cut from the same specimen. The section consists of silicified coniferous wood containing patches of jet. Annual rings are faintly shown, as in 51,487. The jet shows several folded and contorted dark bands, as in Figs. 5 and 7, the general direction of which is parallel to the rows of tracheids in the unaltered patches of wood. The portions converted into jet afford evidence, in the zigzag course of the dark bands, of considerable contraction.

51,593, 51,594. Pure jet similar to that shown in Pl. VIII. Fig. 3, but in this case the arrangement of the longer dark bands across the fainter vertical lines shows that the section has been cut in a radial longitudinal direction.

51,596. Radial longitudinal section of jet, but darker and less transparent than 51,593 and 51,594.

51,597. Probably a tangential longitudinal section, similar to that shown in Fig. 3, Pl. VIII.

51,598. Longitudinal radial section showing the tracheal walls as vertical lines crossed at right angles by rows of dark dots representing medullary-ray cells.

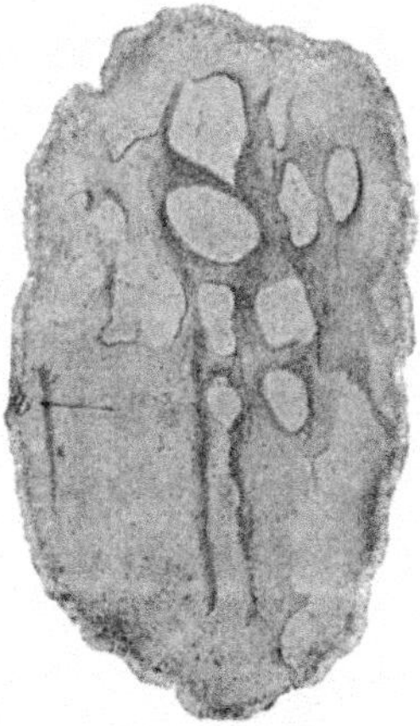

FIG. 7.—Coniferous wood partly converted into jet. Highly magnified. 51,448.

51,599. Transverse section like that reproduced in Text-fig. 7, but darker.

51,600. A good example of pure jet in transverse section; the rich sherry-coloured matrix is traversed by parallel contorted bands, as in Pl. VIII. Fig. 7, and shows traces of the tracheal walls as a faint network of fine lines.

51,621. A small section of jet almost entirely without signs of structure. Under a high power it presents the appearance of a finely and irregularly corrugated surface; this is no doubt a faint reflection of the original structure of the wood as seen in transverse section.

51,622. Dark transverse section of jet. Cf. 51,623 (Fig. 7, Pl. VIII.).

51,782. A small piece of a longitudinal section of jet showing numerous fine dark lines indicating the walls of the tracheids.

51,806. Probably a longitudinal tangential section; the matrix is traversed by wavy dark bands of unequal thickness.

52,623. Specimens of fossil wood and jet.

IV.—OOLITIC SERIES.

The Jurassic system, excluding the Liassic series, may be classified as follows :—

Purbeckian.
Portlandian.
Kimeridgian.
Corallian.
Oxfordian.
Bathonian.
Bajocian.

We are not at present concerned with the Purbeckian and Portlandian fossils; the rest of the series includes several sub-divisions, the nomenclature of which varies to some extent in different parts of Britain. We may, however, adopt the following classification [1] as convenient for reference in connection with the fossils described in this volume.

Groups or Stages.	*Sub-Groups or Sub-Stages.*
KIMERIDGIAN.	Kimeridge Clay.
CORALLIAN.	Coral Rag. Coralline Oolite. Calcareous Grit.
OXFORDIAN.	Oxford Clay. Kellaways Rock.
BATHONIAN (Great or Bath Oolite).	Cornbrash. Bradford Clay and Forest Marble. Great or Bath Oolite, with Stonesfield Slate.
	Fuller's Earth.
BAJOCIAN (Inferior Oolite).	Cheltenham Beds, Lincolnshire Limestones, and Collyweston Slate. Northampton Sands. Midford Sands (passage beds).

[1] Geikie (93).

A complete classification of the Oolitic series should include the various local variations in the nature of the sediments to which special names have been applied. For a more detailed list of the subdivisions of the Oolitic rocks, as founded on the development of the series in different parts of the country, reference should be made to Mr. H. B. Woodward's standard work on the Lower Oolitic rocks of Britain,[1] and to the memoir by the same author on the Middle and Upper Oolitic rocks of England (Yorkshire excepted).[2]

The British Museum collections include fossil plants from several of the above groups; in most cases the specimens from any one area or sub-stage represent mere waifs and strays of the floras, but the Stonesfield Slate (Bathonian) has afforded a comparatively rich collection. In the following pages I have arranged the species botanically, mentioning in each case the locality and horizon from which they have been obtained. The list of plants given in the concluding section serves to indicate the species described from the various sub-stages of the Oolitic rocks.

The Geological Survey memoir on the Jurassic rocks of Britain (vol. iv.) by Mr. H. B. Woodward[3] contains several references to the occurrence of plants in the Lower Oolitic rocks in other parts of England than Yorkshire. Vol. v.[4] of the series of Jurassic memoirs by the same author affords valuable information as to the plant-bearing rocks of the Middle and Upper Oolitic strata; the Yorkshire area is dealt with in a separate volume by Mr. Fox-Strangways.[5]

Several Oolitic plants are figured by Lindley & Hutton in their *Fossil Flora*;[6] these are given in the following list, with the addition of the names employed for their species in the present volume.

---

[1] Woodward (94).
[2] Woodward (95).
[3] Woodward (94).
[4] Woodward (95).
[5] Fox-Strangways (92).
[6] Lindley & Hutton (31–37).

Pinus primæva, L. & H. Pl. cxxxv. 1834. Near Towcester. Inferior Oolite.	= *Conites primæva* (L. & H.).
Zamia pectinata, Brongn. Pl. clxxii. 1835. Stonesfield Slate. Type-specimen in the Oxford Museum.	= *Williamsonia pecten* (Phill.).
Zamia taxina, L. & H. Pl. clxxv. 1835. Stonesfield Slate.	= *W. pecten* (Phill.).
Sphenopteris cysteoides, L. & H. Pl. clxxvi*a*. Stonesfield Slate.	= *Sphenopteris*, sp. *a*.
Tæniopteris vittata, Brongn. Pl. clxxvi*b*. Stonesfield Slate.	= *Tæniopteris vittata*, Brongn.
Carpolithes conica, L. & H. Pl. clxxxix. figs. 1, 2, 4. Coralline Oolite, Malton. Type-specimen (L. & H., fig. 4) in the Manchester Museum, No. 360.	= *Carpolithes conicus*, L. & H.
Carpolithes Bucklandi, L. & H., *ex* Will. MS. Pl. clxxxix. figs. 3, 5. Coralline Oolite, Malton. [A specimen in the Manchester Museum, No. 361, may possibly be the original of fig. 5.]	= *Carpolithes conicus*, L. & H.
Carpolithes. Pl. cxciii*a*, figs. 1–4. Stonesfield Slate.	= *Carpolithes*, sp., and *Araucarites*, sp.

In addition to the plants included in the above list there are others described in more recent years, but we possess no systematic account of Oolitic plants as a whole. Mr. Carruthers has added considerably to our knowledge of Jurassic plants by his numerous papers on Oolitic species from various localities, but reference is made to the work of this author under the head of the species which he has described.

In Damon's *Geology of Weymouth*[1] we find a few references to fossil plants from the Kimeridge Clay and other horizons. In Strahan's Survey memoir allusion is made to the so-called Kimeridge coal, a highly bituminous shaly substance from which 'coal-money' was formerly manufactured.[2] Reference should also

---

1 Damon (84).

2 Strahan (98); see also Buckland & De la Beche (36), and an article in the *Gentleman's Magazine*, vol. xxxviii. p. 111, 1768.

be made to an address delivered by Mr. Etheridge to the Geological Society in 1882, in which several allusions are made to the occurrence of Oolitic plants.[1]

The microscopical investigation of oolitic grains from various Jurassic rocks has raised a question of considerable botanical interest. Wethered[2] and other authors have demonstrated the presence in oolitic grains of narrow tubular elements, referred to the genus *Girvanella*, possibly representing the remains of Algæ which have been instrumental in the production of this common lithological character.

INFERIOR OOLITE (BAJOCIAN) PLANTS.

In a paper *On some Sections in the Oolite District of Lincolnshire*, Morris incidentally notes the common occurrence of the fern *Pecopteris polypodioides*, Brongn. (= *Laccopteris polypodioides*), near Stamford; the same author refers also to the specimens of *Pterophyllum* from the famous Barnack quarries.[3] The abundance of *Laccopteris* in the Collyweston quarries and elsewhere was pointed out by Boscawen, Ibbetson, & Morris in a note communicated to the British Association at the Oxford meeting of 1847.[4] Sharp's papers on *The Oolites of Northamptonshire*[5] also contain references to a few fossil plants from Kingsthorpe and other places. Species of Inferior Oolite plants are mentioned by Professor Judd in his memoir on the Geology of Rutland.[6]

GREAT OOLITE (BATHONIAN) PLANTS.

The Stonesfield Slate is by far the most important sub-stage of the Oolitic series from a botanical standpoint, excluding the Inferior Oolite rocks of Yorkshire. The lowest zone of the Great Oolite series in Oxfordshire and Gloucestershire takes its name from

---

[1] Etheridge (82).

[2] Wethered (89), (90), (95); Harris (96); Seward (98), p. 124, (94^2); Rothpletz (91).

[3] Morris (53), p. 337.

[4] Boscawen, Ibbetson, & Morris (48).

[5] Sharp (70), pp. 361, 384; (73), pp. 274, 290, 295.

[6] Judd (75), pp. 140, 165, 276.

Stonesfield in Oxfordshire, where a fine-grained shelly sandstone has been quarried for roofing-slates for many centuries. In his *Natural History of Oxfordshire*, Plot[1] makes the following reference to the slates of Stonesfield:—"But before we take leave of materials for building, we must not forget that the houses are covered, for the most part in Oxfordshire not with tiles but flat-stone, whereof the lightest, and that which imbibes the water least, is accounted the best. And such is that which they have at Stonesfield, where it is dug first in thick cakes, about Michaelmas time, or before, to lye all the Winter and receive the frosts, which make it cleave in the Spring following into thin slates, which otherwise it would not do so kindly."

The fossil plants of the Stonesfield Slate have been found chiefly at Stonesfield, Sevenhampton Common, and Eyeford; they occur in a marine sediment rich in the remains of mammals, reptiles, insects, and molluscs. It has been suggested that the Stonesfield flora may represent the fragmentary relics of the vegetation of an islet in the Jurassic sea. The frequent occurrence of detached leaflets and the fragmentary condition of the plants may perhaps be regarded as evidence in favour of the view that the area of deposition was a considerable distance from the land on which the vegetation flourished; or it may be that the land rose steeply from the water's edge, and the branches, leaves, and seeds were thus exposed to rough usage by the rivers that swept them into the sea.

In Buckland's paper on the *Megalosaurus or Great Lizard of Stonesfield*, the plants are mentioned but not described.[2] Brongniart, in his *Prodrome*, speaks of the Stonesfield plants as being of rather later date than those from the Yorkshire coast, but he recognises the close agreement between the two floras.[3]

In Murchison's *Outline of the Geology of the Neighbourhood of Cheltenham*, published in 1834,[4] no account of the plants is included; but in a new edition "augmented and revised by James Buckman and H. E. Strickland,"[5] the following species are

---

[1] Plot (1677).
[2] Buckland (24), p. 392.
[3] Brongniart (28), p. 197.
[4] Murchison (34).
[5] Murchison, Buckman, and Strickland (45).

described, several of them being represented by rather crude and inaccurate drawings :—

Bensonia ovata, Buckman. Sevenhampton Common.	= ? *Podozamites stonesfieldensis*, sp. nov.
Cycadites ? Pl. i. fig. 3. Sevenhampton Common.	= *Ctenis latifolia* (Brongn.).
Carpolithus conicus, L. & H. Pl. ii. figs. *5a–d* and *f*, 6. Sevenhampton and Eyeford quarries.	= *Carpolithes conicus*, L. & H.
Thuytes cupressiformis, Sternb. Sevenhampton and Eyeford quarries.	= *Thuites expansus*, Sternb.
T. expansus, Phill. Pl. i. fig. 6. Sevenhampton.	= *T. expansus*, Sternb.
Pinites (?) or other Coniferæ, two species. Sevenhampton Common. *Calamites* (?). Sevenhampton and Eyeford.	I am unable to recognise the specimens so described by Buckman.
Lilia lanceolata, Buck. Pl. ii. fig. 3. Eyeford.	= *Podozamites stonesfieldensis*, sp. nov.
Naiadea obtusa, Buck. Pl. i. fig. 2. Sevenhampton.	= *Sphenozamites Belli*, sp. nov.
N. ovata, Buck. Pl. ii. fig. 1. Sevenhampton.	= *Podozamites stonesfieldensis*, sp. nov.
Noeggerathia (?). Pl. i. fig. 5. Eyeford.	= *Ginkgo digitata* (Brongn.).
Stricklandia acuminata, Buck. Pl. ii. fig. 2. Sevenhampton.	= *Baiera Phillipsi*, Nath.
Salicites longifolius, Buck. Pl. i. fig. 1. Eyeford.	= ? *Thinnfeldia speciosa*, Ett.
Filicites (?). Two species. Eyeford. Musci ?	The specimens so described have not been identified.

In a paper by Brodie & Buckman [1] published in the Journal of the Geological Society in 1845, the Stonesfield plants are briefly noticed. The variable character of the Stonesfield series is emphasised by Hull [2] in the Survey memoir on the country near Cheltenham. Horton,[3] writing on the geology of the Stonesfield Slate in 1860, suggested that the rock may have been deposited in a lagoon bordered by marshes.

---

[1] Brodie & Buckman (45), p. 223.
[2] Hull (57).
[3] Horton (60).

A more complete account of the Stonesfield fossils is given by Phillips in his *Geology of Oxford and the Valley of the Thames*. He speaks of the "Stonesfield lagoon, full of fishes and molluscs, receiving with every cyclonic storm drifted branches of cypresses and swarms of wind-wrecked insects . . . ."[1] As stated in the following list, several of the specimens figured by Phillips are now in the Oxford Museum.

Halymenites ramulosus, Brongn. Stonesfield. Type-specimen, figured by Brongniart in the Hist. vég. foss. pl. iii. fig. 2, doubtfully identified with one in the Oxford Museum.	= ? *Algites furcatus* (Brongn.).
Cyclopteris latifolia, Phill. Eyeford.	= *Ginkgo digitata* (Brongn.).
Glossopteris longifolius, Buck. Eyeford.	= ? cf. *Thinnfeldia speciosa*, Ett.
Hymenopteris macrophylla, Brongn. Stonesfield. Brongniart's type-specimen (Hist. vég. foss. pl. lviii. fig. 3) is in the Oxford Museum.	= cf. *Thinnfeldia speciosa*, Ett.
Pecopteris approximata, Phill. Phillips, Diag. xxviii. fig. 2. Stonesfield.	= *Sphenopteris*, sp. *b*.
P. diversa, Phill. Diag. xxviii. fig. 1. Stonesfield. Type-specimen (a very imperfect and obscure fragment) in the Oxford Museum.	= ? *Nageiopsis*, sp.
P. incisa, Phill. Diag. xxviii. figs. 5–6. Stonesfield. Type-specimen (indistinctly preserved fragment) in the Oxford Museum.	= *Sphenopteris*, sp. *b*.
Sphenopteris cysteoides, L. & H. Phillips, Diag. xxviii. fig. 7. Stonesfield. Type-specimen in the Oxford Museum.	= *Sphenopteris*, sp. *a*.
S. plumosa, Phill. Diag. xxviii. figs. 3–4. Stonesfield.	= *Sphenopteris*, sp. *b*.
Tæniopteris latifolia, Brongn. Stonesfield. Type-specimen of Brongniart (Hist. vég. foss. pl. lxxxii. fig. 6) in the Oxford Museum.	= *Ctenis latifolia* (Brongn.).

[1] Phillips (71), p. 237.

Tæniopteris angustata, Phill. Diag. xxviii. figs. 8–10. Stonesfield.	= *Laccopteris Woodwardi* (Leck.).
T. scitamineæ-folia, Sternb. Phillips, Diag. xxx. fig. 8. Stonesfield. Sternberg's type-specimen, figured also by Phillips, in the Oxford Museum.	= *Tæniopteris vittata*, Brongn.
Aroides Stutterdi, Carr. Stonesfield. Figured by Carruthers in the Geol. Mag. vol. iv. pl. viii. 1867. Phillips, Diag. xxxii. figs. 12–13.	I am unable to name this fossil, and do not regard it as a plant.
Bensonia ovata, Buck. Sevenhampton. *Lilia lanceolata*, Buck. Eyeford. No figures are given of these species, but Phillips describes them as perhaps Cycads allied to *Dioon*.	= *Podozamites stonesfieldensis*, sp. nov.
Naiadea obtusa, Buck. Sevenhampton.	= *Sphenozamites Belli*, sp. nov.
N. ovata, Buck. Sevenhampton and Eyeford.	= *Podozamites stonesfieldensis*, sp. nov.
Stricklandia acuminata, Buck. Sevenhampton.	= *Baiera Phillipsi*, Nath.
Bucklandia squamosa, Sternb. Stonesfield. Type - specimen of Sternberg, figured also by Phillips (Diag. xxix.), in the Oxford Museum.	= *Cycadeoidea squamosa* (Sternb.).
Palæozamia longifolia, Phill. Diag. xxx. fig. 6. Stonesfield. Type-specimen in the Oxford Museum. (Refigured in the present volume, Fig. 11, p. 114.)	= *Zamites megaphyllus* (Phill.).
P. megaphylla, Phill. Diag. xxx. fig. 1. Stonesfield. Type - specimen in the Oxford Museum.	= *Z. megaphyllus* (Phill.).
P. pecten, Phill. Stonesfield. *P. pectinata* (Brongn.). Diag. xxx. fig. 2. Original in the Oxford Museum. The specimen from Stonesfield figured by Lindley & Hutton (pl. 172) is also in the Oxford Museum.	= *Williamsonia pecten* (Phill.).
P. taxina, L. & H. Phillips, Diag. xxx. figs. 4–5. Stonesfield.	= *W. pecten* (Phill.).

Pterophyllum Buckmanni, Phill. Sevenhampton. *P. comptum*, Phill. Stonesfield. *P. minus* (?), Brongn. Stonesfield.	I am unable to identify with certainty the species referred to by Phillips under these names.
Taxites podocarpioides, Brongn. Diag. xxxi. fig. 6; Diag. xxx. fig. 7. Stonesfield. Type-specimen of Brongniart in the Oxford Museum (Hist. vég. foss. pl. lviii. fig. 3), figured also by Phillips.	= Cf. *Thinnfeldia speciosa*, Ett.
Thuyites articulatus, Sternb. Phillips, Diag. xxxi. fig. 1. Stonesfield.	= ? *Thuites expansus*, Sternb.
T. cupressiformis, Sternb. Stonesfield. *T. divaricatus*, Sternb. Stonesfield. *T. expansus*, Sternb. Phillips, Diag. xxxi. fig. 5. Stonesfield.	= *Thuites expansus*, Sternb.
Carpolithus diospyriformis, Sternb. Phillips, Diag. xxxii. fig. 2. Stonesfield.	= *Carpolithes diospyriformis*, Sternb.
C. Lindleyanus, Gutb. Phillips, Diag. xxxii. fig. 1. Stonesfield.	= ? *Araucarites*, sp.

Very few fossil plants are recorded from the other sub-stages of the Great Oolite series, and it is unnecessary to refer to the small number of records to be found in the literature to Bathonian plants other than those from the Stonesfield Slate.

Oxfordian Plants.—In addition to fossil wood,[1] a very few species of plants are known from the Oxford Clay. From Wiltshire, Carruthers[2] has described a fossil which he named *Cycadeostrobus sphæricus*, but this is in all probability an Araucarian cone. The Museum collection contains pieces of coniferous twigs from Christian Malford, some of which have been described by Carruthers.[3]

---

[1] Porter (63).
[2] Carruthers (67).
[3] Carruthers (69), p. 7.

Corallian Plants.—A small number of plants have been obtained from the Coralline Oolite of Malton in Yorkshire; the most abundant specimens are the Gymnospermous seeds described by Lindley & Hutton [1] and other writers. The Malton Museum contains some good examples; others may be seen in the British Museum, the Manchester Museum, the York Museum, the Museum of Practical Geology, London, and in several other collections. Reference is made to Malton plants by Blake & Hudleston [2] in their paper on the Corallian rocks of England, and in an appendix Carruthers [3] describes an interesting example of an Araucarian cone, *Araucarites Hudlestoni*, preserved in the York Museum.

Kimeridgian Plants.—Reference has already been made to the Kimeridge Coal. In 1869 Carruthers [4] described a small cone from the Kimeridge Clay of Weymouth, which he named *Pinites depressus*; the type-specimen, which has almost fallen to pieces, is in the British Museum (V. 6370). In 1892 Mr. George Murray [5] proposed the name *Caulerpa Carruthersi* for some specimens which he believed to represent a species of the Siphoneous algal genus *Caulerpa*, but, as pointed out in the sequel, it is probable that the specimens so named do not owe their origin to algæ.

Class THALLOPHYTA.

The British Oolitic rocks, at least those which are dealt with in these pages, have not afforded any thoroughly satisfactory examples of either Algæ or Fungi. In the petrified tissues of coniferous wood one constantly finds direct or indirect evidence of the occurrence of fungi, but I am not aware of any examples of this group worthy of record among the fossils in the Museum collections. Several 'species' of Algæ have been described, but none have any claim to be regarded as botanically important.

---

1 Lindley & Hutton (31–37).
2 Blake & Hudleston (77), p. 367.
3 Carruthers (77), p. 402, pl. xvii.
4 Carruthers (69), pl. ii. fig. 10.
5 Murray (92).

Genus **ALGITES**, Seward.

[Wealden Flora, vol. i. p. 4, 1894.]

? **Algites furcatus** (Brongn.).

[Prodrome, p. 198, 1828.]

1828. *Fucoides furcatus*, Brongniart, Prodrome, p. 198.
F. furcatus, Brongniart, Hist. vég. foss. p. 62, pl. iii. fig. 1.
1838. *Halymenites ramulosus*, Sternberg, Flor. Vorwelt, vii. p. 31.
1845. *H. ramulosus*, Unger, Synopsis, p. 13.
1848. *H. ramulosus*, Bronn, Ind. Pal. p. 565.
1849. *Sphærococcites ramulosus*, Brongniart, Tableau, p. 105.
1850. *Halymenites ramulosus*, Unger, Gen. spec. foss. plant. p. 23.
1854. *H. ramulosus*, Morris, Brit. Foss. p. 8.
1856. *H. ramulosus*, Zigno, Flor. foss. Oolit. vol. i. p. 21.
1871. *H. ramulosus*, Phillips, Geol. Oxford, p. 168.
1894. *H. ramulosus*, Woodward, Lower Ool. p. 600.

The specimen from Stonesfield to which Brongniart gave the name *Fucoides furcatus* is preserved in the Oxford Museum. It is too obscure and imperfect to be identified as an alga with any degree of certainty; like so many of the fossils referred to the Algæ, it represents the impression of a dichotomously branched axis, and bears a resemblance to the forked thallus of *Fucus* and many other genera of recent Algæ. We may retain Brongniart's specific name, with the substitution of the generic term *Algites*, although specimens of this kind are valueless, and it serves no useful purpose to include them in lists of fossil plants.

I have seen no specimens, with the exception of that in the Oxford Museum, which can be referred to Brongniart's species.

[*Fucoides elegans*, Brongniart. This name was applied by Brongniart in 1823 to a specimen from Stonesfield of which a drawing was sent to him by Buckland. In his later work (Hist. vég. foss. 1828) Brongniart points out that an inspection of the fossil itself in the Oxford Museum convinced him that it should be referred to the Coniferæ, and not to the Algæ. The specimen is no doubt a fragment of a coniferous twig, but it is too small and incomplete to designate by a specific name.]

Genus **GIRVANELLA**, Nicholson & Etheridge.

[Silurian Foss., Girvan District, p. 23, 1880.]

Girvanella pisolitica, Wethered.

[Geol. Mag. (iii.), vol. vi. p. 200, pl. vi. figs. 8, 10, 11, 1889.]

1889. *Girvanella pisolitica*, Wethered, Geol. Mag. [iii.], vol. vi. p. 200, pl. vi. figs. 8, 10, 11.

1891. *G. pisolitica*, Rothpletz, Zeitschr. deutsch. geol. Ges. vol. xliii. p. 295.

Mr. Wethered describes the species as follows:—"This species occurs in the form of flattened spherules varying in size from 1 to $\frac{5}{12}$ inch in greatest diameter. In the centre of each spherule there is a nucleus which is surrounded by calcareous tubuli with well-defined walls, and averaging about $\frac{1}{500}$ of an inch in diameter, though some are smaller. In some instances, more especially in the larger spherules, the tubes bend and twist about in a truly vermiform manner, often assuming the form of a flattened coil . . . . The tubes are also branching, and are larger than those of *G. problematica*."

P. 4481. Two sections of oolitic grains containing *G. pisolitica* from Cleeve Hill, Gloucestershire (Inferior Oolite).

Presented by E. Wethered, Esq., 1892.

The wide geological range of *Girvanella*, from Cambrian rocks upwards, and the very small knowledge we possess as to the nature of the genus, render it inadvisable to attempt a specific diagnosis and terminology for the numerous forms of tubules met with in Oolitic rocks. Rothpletz [1] and Brown [2] have expressed the opinion that *Girvanella* is probably a calcareous alga of the group Siphoneæ. I have elsewhere suggested that the tubules may be the sheaths of Cyanophyceæ, but we require more information before we can hope to speak with confidence as to the systematic position of the genus.[3]

In addition to the type named by Mr. Wethered *Girvanella pisolitica*, other species have been described by the same author from Jurassic Oolites, but none of these are represented in the Museum collection.

---

[1] Rothpletz (91), p. 301.
[2] Brown (94), p. 203.
[3] Seward (98), p. 125.

Genus **SOLENOPORA**, Dybowski.

[Die Chætetiden der Osbalt. Sil. Form. p. 124, pl. ii. 1877.]

Solenopora jurassica, Brown *ex* Nicholson MS.

[Geol. Mag. (iv.), vol. i. p. 150, figs. 4–5, 1894.]

Dr. Alexander Brown[1] published an interesting account of the structure of several forms of calcareous organisms from Palæozoic and Mesozoic rocks, which had been classed by previous writers among the Hydrozoa. He describes a species from the Great Oolite of Chedworth, Gloucestershire, and Malton, Yorkshire, under the name *Solenopora jurassica*, Brown *ex* Nich. MS. As Dr. Brown has shown, there are good reasons for assigning *Solenopora* to the Corallineæ, a family of the Florideæ, to which the well-known *Lithothamnion* and other existing types belong.

Caulerpa Carruthersi, Murray.

[Phycolog. Memoirs, pt. i. p. 11, 1892.]

1888. "Equisetaceous plant," Damon, Geol. Weymouth, pl. xix. fig. 12.
1892. *Caulerpa Carruthersi*, Murray, Phycolog. Mem. pt. i. p. 11, pls. iv.–v.
1895. *C. Carruthersi*, Woodward, Lower Ool. p. 402.

As I have elsewhere[2] discussed the nature of the Kimeridge fossils referred to *Caulerpa*, it is needless to traverse the same ground again. It is, I believe, highly improbable that the specimens represent the remains of an alga. The best specimens have the form of a slender central axis, giving off at fairly regular intervals whorls of short and somewhat clavate branches; they bear a superficial resemblance to such a recent species as *Caulerpa cactoides*, Ag. We must leave these fossils as indeterminable, with the suggestion that they have probably been produced by animal—rather than by plant—agency.

The specimens figured by Mr. Murray are in the Botanical Gallery of the British Museum and in the Museum of Practical

---

[1] Brown (94), p. 150, figs. 4–5. See also Seward (98), p. 189.
[2] Seward (94), p. 2; (98), p. 158.

Geology, Jermyn Street. The following examples are in the Geological Department (Fossil Plant Gallery):—

V. 2546a. Figured by Damon[1] as an Equisetaceous plant.

Kimeridge Clay, Sandsfoot, Dorsetshire. *Damon Coll.*

Other specimens:—**V. 25** (several examples), Kimeridge Clay, Weymouth; **V. 2546**, Sandsfoot; **52,529**, Weymouth.

Damon Coll. and *purchased*, 1870.

Group EQUISETALES.

In Buckman's list of Stonesfield fossils published in the *Geology of Cheltenham* the name "*Calamites*?" occurs as a designation, applied with doubt, to certain specimens from Sevenhampton and Eyeford. It is impossible to express any opinion as to the Equisetaceous nature of the fossils, as neither a figure nor a description is given. I have not seen a single specimen among the large number of British Oolitic plants (excluding those from the Yorkshire coast) that can confidently be referred to the Equisetales.

Group FILICALES.

Family MATONINEÆ.

The two existing ferns *Matonia pectinata*, R. Br., and the species *M. sarmentosa*, Baker, are usually placed in a separate division of Leptosporangiate ferns as survivals of a family which was widely distributed during the Rhætic and Jurassic periods. The two fossil genera *Matonidium* and *Laccopteris* agree very closely, both in their habit and as regards the soral characters, with the Malayan genus *Matonia*. We may better express the peculiarities and isolated position of these ferns by assigning them to a family apart than by including *Matonia*, with its fossil allies, in the large division of Polypodiaceæ. As our knowledge of the morphological characters of the comprehensive family Polypodiaceæ becomes more complete, it seems clear that under that title are included genera which cannot be regarded as constituting a natural family.

---

[1] Damon (88), pl. xix. fig. 12.

Genus LACCOPTERIS, Presl.

[Presl, in Sternberg, Flor. Vorwelt, vii. p. 115, 1838.]

Laccopteris Woodwardi (Leckenby).

[Quart. Journ. Geol. Soc. vol. xx. p. 81, 1864.]

1848. *Pecopteris polypodioides*, Ibbetson & Morris, Brit. Assoc. Rep. p. 128.
1853. *P. polypodioides*, Morris, Quart. Journ. Geol. Soc. vol. ix. p. 337.
1856. *Phlebopteris Woodwardi*, Zigno, Flor. foss. Oolit. vol. i. p. 174.
1864. *P. Woodwardi*, Leckenby, Quart. Journ. Geol. Soc. vol. **xx**. p. 81, pl. viii. fig. 6.
1869. *P. Woodwardii*, Schimper, Trait. pal. vég. vol. i. p. 626.
1871. *Tæniopteris angustata*, Phillips, Geol. Oxford, p. 168.
1873. *Microdictyon Woodwardianum*, Saporta, Pal. Franç. vol. i. p. 313, pl. xxxiii.
M. rutenicum, ibid. pl. xxxiii. figs. 2–4; pl. xliv.
Pecopteris polypodioides, Sharp, Quart. Journ. Geol. Soc. vol. **xxix**. p. 295.
1875. *Phlebopteris Woodwardii*, Phillips, Geol. Yorks. p. 202.
Pecopteris polypodioides, Judd, Geol. Rutland, p. 140.
1892. *Phlebopteris Woodwardi*, Fox-Strangways, Tab. Foss. p. 134.
Cf. *Microdictyon Woodwardianum*, Bartholin, Bot. Tid. Kjövenhavn, p. 24, pl. x. figs. 2–4.
1894. *Tæniopteris angustata*, Woodward, Lower Ool. p. 317.
Phlebopteris polypodioides, ibid. p. 600.
1899. *Laccopteris Woodwardi*, Seward, Trans. R. Soc. vol. 191, p. 198, fig. 9*a*.
1900. *L. Woodwardi*, Seward, Jurassic Flora, vol. i. p. 84, fig. 11*a*.
1902. *Microdictyon Woodwardii*, Möller, Bornholms Foss. Flor. i. pl. iv. figs. 2–4.

Leckenby's type-specimen is in the Sedgwick Geological Museum, Cambridge (No. 126). The fragments from Stonesfield figured by Phillips in his *Geology of Oxford*[1] (Diag. xxviii. figs. 8–10) as *Tæniopteris angustata*, Phill., are in the Oxford Museum.

Localities and Horizons. — Gristhorpe Bay, Yorkshire coast (Inferior Oolite). Stamford, Lincolnshire; Rothwell, near Kettering, Northamptonshire [Inferior Oolite (Lincolnshire Limestone)]. Ponton and Collyweston, Lincolnshire [Inferior Oolite (Collyweston Slate)]. Kingsthorpe and Weekley, Northamptonshire (Great Oolite).

---

[1] Phillips (71).

Habit most probably like that of *Laccopteris polypodioides*. Frond digitate, divided into spreading pinnatifid branches with linear ultimate segments. Ultimate segments traversed by a prominent midrib giving off secondary veins at a wide angle; these veins anastomose and form a series of comparatively broad meshes on each side of the midrib. From the outer edge of these meshes arise numerous tertiary veins joined to one another by oblique cross-connections. Sori circular, probably without an indusium, borne in two rows, one on each side of the central rib.

This species is represented by an abundance of pinnule fragments in the Inferior Oolite of Stamford and other places. It is impossible to give a complete diagnosis of this type, but it agrees closely with *Laccopteris polypodioides*,[1] Brongn., except in the more numerous tertiary veins in the ultimate segments; the sori are in two rows parallel to the midrib, and of circular form. In habit the fronds were no doubt like those of *Laccopteris polypodioides*, and very similar to those of the recent *Matonia pectinata*. In all probability some fragments from the Perucer Beds of Bohemia recently figured by Frič & Bayer[2] as *Drynaria* should be referred to *Laccopteris*.

One of the specimens (20,040*d*) in the Museum collection includes a fragment of what is probably the rachis, showing the form of the petiolar vascular strand, which agrees with that of *Matonia* in being broadly U-shaped, with the ends of the arms bent inwards at right angles. In another specimen the edges of the lamina of the pinnules are seen to be strongly revolute.

The Jermyn Street Museum collection includes some good specimens from Stamford, and in the Northampton Museum I have seen fragments of the same species from the Lincolnshire Limestone of Rothwell, near Kettering.

V. 4646. Fertile pinnules.

Collyweston Slate. *C. W. Peach Coll.*

20,040*c*, 20,040*d*. Fragments of pinnules; a section of a rachis (?) seen in 20,040*d*.

Weekley, Northants. *Miss Baker's Coll.*

---

[1] Seward (00), p. 78, pls. xii.–xiii.; text-figs. 8–11.

[2] Frič & Bayer (01).

52,548. Numerous fertile pinnules showing the sori, but no individual sporangia; edges of pinnule revolute as in *Matonia pectinata*.

Collyweston. *Morris Coll.*

52,870. Several fragments of fertile pinnules crowded together as if deposited in an eddy; circular sori clearly shown.

Ponton, Lincs. *Sharp Coll.*

Other specimens:—**52,584, 52,869, 82,439** (Stamford).

Sharp & Etheridge Colls.

Family DIPTERIDINÆ.

In this family of Leptosporangiate ferns are included the recent genus *Dipteris*, represented by four species in India, the Malay Peninsula, and elsewhere, and several Rhætic and later Mesozoic species referred to *Dictyophyllum* and other genera. Like the Matonineæ, the Dipteridinæ[1] must be looked upon as survivals from the Mesozoic era, when they were widely spread throughout Europe and constituted prominent members of Rhætic and Jurassic floras.

Genus **DICTYOPHYLLUM**, Lindley & Hutton.

[Foss. Flor. ii. pl. civ. 1834.]

It is often difficult to decide between the generic names *Dictyophyllum* and *Protorhipis* in naming imperfect specimens of fronds, and it is by no means improbable that the leaves referred to both genera may be generically identical. A comparison of various types of *Dipteris* fronds, as shown in the drawings published in a paper on recent and fossil Dipteridinæ by Miss Dale and myself, demonstrates the futility of attempting to discriminate between fragments of Mesozoic leaves by the employment of different generic names. In cases where the material admits of a fairly complete diagnosis the employment of distinctive names may serve a useful purpose, but a specimen such as that represented in Text-fig. 8 is too imperfect to identify within narrow limits.

---

[1] Seward (00), p. 118; Seward & Dale (01), p. 36.

Dictyophyllum, sp.

Localities and Horizons.—Collyweston (Inferior Oolite); Stonesfield ? (Great Oolite).

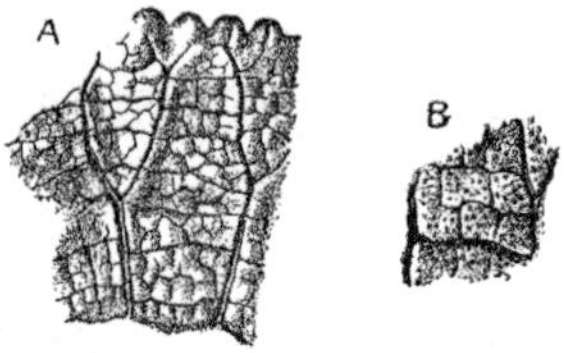

FIG. 8.—*Dictyophyllum*, sp. A, fragment of sterile lamina; B, fragment of fertile lamina. V. 9003. Nat. size.

V. 9003. Text-fig. 8.

An imperfect fragment of a leaf with a bluntly crenulate margin showing the venation-characters of *Dictyophyllum* and *Protorhipis*. The fragment B, Fig. 8, affords slight indications of sori. The larger piece (A) bears a close resemblance to leaves of the genus *Protorhipis*, as for example the type-specimen of Andrae's *P. Buchii*,[1] but the material is too imperfect to identify with certainty. A species from the Lower Jurassic rocks of Cracow,[2] described by Raciborski as *Dictyophyllum cracoviense*, bears a close resemblance to the fragment represented in Text-fig. 8. A comparison may be made also with the small frond of *Dipteris conjugata* figured in pl. xlviii. fig. 23, of the paper[3] on recent and fossil Dipteridinæ already referred to.

Collyweston Slate. *C. W. Peach Coll.*

V. 4672. Fragment of the same type of leaf as that shown in Text-fig. 8.

Collyweston. *C. W. Peach Coll.*

A fragment of *Dictyophyllum* in the Oxford Museum (Stonesfield ?) may be specifically identical with the Collyweston plant.

---

[1] Andrae (53), pl. viii. fig. 1.

[2] Raciborski (94), pl. xiv. figs. 5–10.

[3] Seward & Dale (01).

FERNS OF DOUBTFUL FAMILY POSITION.

Genus **TÆNIOPTERIS**, Brongniart.

[Prodrome, p. 61, 1828.]

Tæniopteris vittata, Brongniart.

[Hist. vég. foss. p. 263, pl. lxxxii. figs. 1–4, 1828.]

1821. *Tæniopteris scitaminea*, Sternberg, Flor. Vorwelt, ii. p. 140.
1823. *Scitaminearum folium*, op. cit. iii. p. 42, pl. xxxvii. fig. 2.
1828. *Tæniopteris vittata*, Brongniart, Prodrome, p. 62.
T. vittata, Brongniart, Hist. vég. foss. p. 263, pl. lxxxii. figs. 1–4.
1835. *T. vittata*, Lindley & Hutton, Foss. Flor. vol. iii. pl. clxxvi *b*.
1838. *T. scitaminea*, Sternberg, Flor. Vorwelt, vii. p. 139.
1845. *T. scitaminea*, Unger, Synopsis, p. 37.
1848. *T. scitaminea*, Bronn, Ind. Pal. p. 1215.
1849. *T. vittata*, Brongniart, Tableau, p. 105.
1850. *T. scitaminea*, Unger, Gen. spec. foss. p. 212.
1854. *T. scitaminea*, Morris, Brit. Foss. p. 23.
1856. *T. vittata*, Zigno, Flor. foss. Oolit. vol. i. p. 201.
Stangerites vittatus, Bornemann, Organ. Rest. Lettenkohle Thüring. p. 60.
1869. *Oleandridium vittatum*, Schimper, Trait. pal. vég. vol. i. p. 608.
1871. *Tæniopteris scitamineæ-folia*, Phillips, Oxford, p. 168, Diag. xxx. fig. 8.
1873. *T. vittata*, Saporta, Pal. Franç. vol. i. p. 444, pl. lxiv. figs. 1–5.
1894. *T. vittata*, Woodward, Lower Ool. p. 600.
1902. ? *T. tenuinervis*, Möller, Bornholms Foss. Flor. i. p. 37, pl. iii. figs. 12–16.

Locality and Horizon.—Stonesfield: Stonesfield Slate.

The specimen figured by Phillips (Geol. Oxford, Diag. xxx. fig. 8) is in the Oxford Museum. A figure of the same specimen is given also by Sternberg (Flor. Vorwelt, pl. xxxvii. fig. 2, 1823). The synonymy of this species is given in vol. i. of the *Jurassic Flora*,[1] and need not be repeated here in detail. The fragment of a frond figured by Sternberg[2] and afterwards by Phillips[3] as *Tæniopteris scitaminea* from the Stonesfield Slate is in the Oxford Museum; it represents the apical portion of a leaf with a strong midrib giving off slightly oblique secondary veins. Sternberg's

---

[1] Seward (00), p. 156.
[2] Sternberg (23), pl. xxxvii. fig. 2.
[3] Phillips (71), p. 168.

species has been regarded by some authors[1] as identical with Brongniart's *Tæniopteris vittata*, and, so far as the small specimen enables one to form an opinion, I believe the Stonesfield Slate fragment to be specifically identical with the Yorkshire Coast fern, *Tæniopteris vittata*. No specimens of *Tæniopteris* have been found among the Stonesfield Slate plants in the British Museum.

Genus **SPHENOPTERIS**, Brongniart.

[Mém. Mus. Hist. nat. Paris, vol. iii. p. 233, 1822.]

Lindley & Hutton have described a fragment of a Stonesfield fern as *Sphenopteris cysteoides*; the same species is figured also by Phillips, and this author has referred other fragments to *Sphenopteris plumosa* and to *Pecopteris approximata* and *P. incisa*.

None of the specimens named by these writers are sufficiently large or well preserved to be made the 'types' of new species.

SPHENOPTERIS, sp. *a*.

[Cf. *Coniopteris quinqueloba* (Phill.), Geol. Yorks. 3rd ed. p. 215, Lign. 33, 1875.]

1835. *Sphenopteris cysteoides*, Lindley & Hutton, Foss. Flor. vol. iii. pl. clxxvi *a*.
1848. *S. cysteoides*, Bronn, Ind. Pal. p. 1168.
1849. *S. cysteoides*, Brongniart, Tableau, p. 105.
1854. *S. cysteoides*, Morris, Brit. Foss. p. 21.
1856. *S. cysteoides*, Zigno, Flor. foss. Oolit. vol. i. p. 79.
1871. *S. cysteoides*, Phillips, Geol. Oxford, p. 168, Diag. xxviii. fig. 7.
1874. *S. cysteoides*, Schimper, Trait. pal. vég. vol. iii. p. 467.
1894. *S. cysteoides*, Woodward, Lower Ool. p. 600.

The fossil figured by Lindley & Hutton from the Stonesfield Slate is described as a ferruginous impression, and, judging by other examples in the Oxford Museum and elsewhere, the type-specimen was most probably too obscure and imperfect to be designated by a specific name. The fragment figured by Phillips is in the Oxford Museum, and is too indistinct to determine, even generically; it may, however, be a fragment of a Conifer. The British Museum collection does not include any specimens that appear to be identical with *Sphenopteris cysteoides*, L. & H.

---

[1] Brongniart (28), p. 62; Unger (50), p. 212; Zigno (56), p. 201.

SPHENOPTERIS, sp. *b*.

[Cf. *Coniopteris hymenophylloides* (Brongn.), Hist. vég. foss. p. 189, pl. lvi. fig. 4, 1828.]

1871. *Sphenopteris plumosa*, Phillips, Geol. Oxford, p. 168, Diag. xxviii. figs. 3–4.
Pecopteris approximata, ibid. Diag. xxviii. fig. 2.
P. incisa, ibid. Diag. xxviii. figs. 5–6.
1894. *Sphenopteris plumosa*, Woodward, Lower Ool. p. 600.
Pecopteris approximata, ibid.
P. incisa, ibid.

The fragment named by Phillips *Sphenopteris plumosa* probably belongs to *Coniopteris hymenophylloides* (Brongn.),[1] an abundant and variable fern in the Inferior Oolite strata of the Yorkshire coast; but in the absence of evidence as to the fertile pinnæ it is safer to adopt the provisional genus *Sphenopteris*.

It should be noted that McCoy has described an Australian fern under the name *S. plumosa*. The type of Phillips' *Pecopteris approximata* is also a fragment too small to be identified; it may perhaps be a portion of a *Coniopteris* pinna. The type-specimen of *Pecopteris incisa* in the Oxford Museum is a badly preserved imperfect fragment, probably specifically identical with the other specimens named by Phillips. Sternberg also made use of the name *P. incisa* for a species from the Coal-measures.

V. 3437. A very indistinct specimen which is probably identical specifically with *Sphenopteris*, sp. *b*.

Eyeford, Stonesfield Slate. *Brodie Coll.*

Genus SAGENOPTERIS, Presl.

[Sternberg, Flor. Vorwelt, vii. p. 164, 1838.]

We have not as yet obtained sufficiently good evidence to justify the inclusion of this genus among the Marsiliaceæ, and I prefer, therefore, to regard *Sagenopteris* as a fern. Its position is by no means settled, and the discovery of well-preserved fertile leaves or sporangia may demonstrate the correctness of Nathorst's view that this Rhætic and Jurassic type is closely allied to the recent *Marsilia*.

---

[1] Seward (00), p. 99, pls. xvi., xvii., xx., xxi.

Sagenopteris Phillipsi (Brongniart).

[Hist. vég. foss. p. 225, pl. lxi. *bis*, fig. 5; pl. lxiii. fig. 2, 1828.]

(Pl. IX. Fig. 3.)

1828. *Glossopteris Phillipsii*, Brongniart, Hist. vég. foss. p. 225, pl. lxi. *bis*, fig. 5; pl. lxiii. fig. 2.
1833. *G. Phillipsii*, Lindley & Hutton, Foss. Flor. pl. lxiii.
1835. *Otopteris cuneata*, ibid. pl. clv.
1836. *Acrostichites Phillipsii*, Göppert, Foss. Farrn. p. 286.
1838. *Sagenopteris Phillipsii*, Sternberg, Flor. Vorwelt, vii. p. 165.
1848. *Tæniopteris Phillipsii*, Bronn, Ind. Pal. p. 1215.
1851. *Sagenopteris cuneata*, Bunbury, Quart. Journ. Geol. Soc. vol. vii. p. 184.
1856. *Phyllopteris Phillipsii*, Zigno, Flor. foss. Oolit. vol. i. p. 166.
1875. *Glossopteris Phillipsii*, Phillips, Geol. Yorks. Coast, p. 203, pl. viii. fig. 8.
1900. *Sagenopteris Phillipsi*, Seward, Manchester Lit. Phil. Soc. vol. xliv. p. 11.
S. Phillipsi, Seward, Jurass. Flor. i. p. 162, pl. xviii. figs. 2–4; text-figs. 24–26.
1902. *S. Phillipsii*, Möller, Bornholms Foss. Flor. p. 52, pl. vi. figs. 1–7.
S. cuneata, ibid. p. 54, pl. vi. fig. 10.

A more complete synonymy is given in vol. i. of the *Jurassic Flora* and by Möller in his recent memoir on the fossil plants of Bornholm. Reference may be made to my previous volume[1] and to a paper published in the Proceedings of the Literary and Philosophical Society of Manchester[2] for a general account of the species and for a discussion on the nomenclature.

V. 4071. Pl. IX. Fig. 3.

A good impression in oxide of iron of two leaflets of asymmetrical form, showing the anastomosing venation characteristic of the genus. The specimen agrees too closely with those from the Yorkshire coast (Inferior Oolite) to be referred to a distinct species.

Stonesfield?

V. 4655, V. 4667. A single leaflet.

The Jermyn Street collection contains a single leaflet from Stonesfield labelled *Pachypteris lanceolata*, which is undoubtedly a fragment of *Sagenopteris*.

---

[1] Seward (00), pp. 163–165.
[2] Seward (00^2), p. 11.

Genus THINNFELDIA, Ettingshausen.

[Abh. k.-k. geol. Reichs. vol. i. Abth. 3, p. 2, 1852.]

Some at least of the imperfect specimens described below are referred without hesitation to the genus *Thinnfeldia*, as represented by *T. speciosa*, Ett.; others exhibit a close agreement with species included by Saporta in his genus *Stenopteris*. The Stonesfield material is very meagre and imperfectly preserved, but on the whole I believe it is better to refer all the specimens to *Thinnfeldia*, limiting the generic name *Stenopteris* to plants with narrow linear ultimate branches in which no lateral veins occur. An example of *Stenopteris*, using the term in this sense, is afforded by a species described by Carruthers from Rhætic beds in the Argentine as *Sphenopteris elongata*,[1] and recorded also from the Stormberg beds of Cape Colony.[2]

Cf. Thinnfeldia speciosa, Ettingshausen.

(Pl. X. Figs. 1–3.)

1828. *Sphenopteris* (?) *macrophylla*, Brongniart, Prodrome, p. 51.
Taxites podocarpoides, Brongniart, ibid. p. 108.
Sphenopteris (?) *macrophylla*, Brongniart, Hist. vég. foss. p. 212, pl. lviii. fig. 3.
1833. *S.* (?) *macrophylla*, Sternberg, Flor. Vorwelt, v.–vi. p. 65.
1836. *Hymenophyllites macrophylla*, Göppert, Syst. Fil. p. 262.
1838. *Rhodea macrophylla*, Sternberg, Flor. Vorwelt, vii. p. 111.
1845. *Hymenophyllites macrophyllus*, Unger, Synopsis, p. 71.
Taxites podocarpoides, ibid. p. 210.
Salicites longifolius, Buckman, Geol. Chelt. p. 68, pl. i. fig. 1.
1848. *Taxites podocarpoides*, Bronn, Ind. Pal. p. 1216.
Hymenophyllites macrophyllus, ibid. p. 602.
1849. *H. macrophyllus*, Brongniart, Tableau, p. 105.
Taxites podocarpoides, ibid. p. 106.
? *Moreauia podocarpoides*, Pomel, Flor. foss. France, p. 22.
1850. *Hymenophyllites macrophyllus*, Unger, Gen. spec. foss. plant. p. 131.
Taxites podocarpoides, ibid. p. 390.
T. podocarpoides, Göppert, Foss. Conif. p. 246.
1852. Cf. *Thinnfeldia speciosa*, Ettingshausen, Abh. k.-k. geol. Reichs. vol. i. Abth. 3, p. 4, pl. i. fig. 8.

---

[1] Carruthers (72), pl. xxvii. fig. 1.
[2] Seward (03), p. 70, pl. vii. figs. 2–3; pl. xi. fig. 3.

1854. *Hymenophyllites macrophyllus*, Morris, Brit. Foss. p. 10.
Taxites podocarpoides, ibid. p. 23.
1856. *Hymenophyllites macrophyllus*, Zigno, Flor. foss. Oolit. vol. i. p. 87.
1866. *Taxites podocarpoides*, Carruthers, Geol. Mag. vol. iii. p. 545.
1871. *Hymenopteris macrophylla*, Phillips, Geol. Oxford, p. 168.
Taxites podocarpoides, ibid. p. 171, Diag. xxxi. fig. 6.
"Ramose plant, *T. podocarpoides*," ibid. Diag. xxx. fig. 7.
Glossopteris longifolius, ibid. p. 168.
1873. *Stenopteris desmomera*, Saporta, Pal. Franç. vol. i. p. 292, pl. xxxii. figs. 1–2; pl. xxxiii. fig. 1.
S. desmomera, Saporta, Plant. foss. Cerin. p. 22, pl. xiv. fig. 2.
1874. *S. desmomera*, Schimper, Trait. pal. vég. vol. iii. p. 511, pl. cvii.
1894. *Hymenophyllites macrophyllus*, Woodward, Lower Ool. p. 599.
Taxites podocarpoides, ibid. p. 599.
Salicites longifolius, ibid. p. 598.

The type-specimen (Stonesfield) of Brongniart's *Sphenopteris macrophyllus* is in the Oxford Museum; it consists of an imperfectly preserved fragment of what appears to be a bipinnate frond with narrow linear ultimate segments. The fragment figured by Phillips as a ramose plant, and referred with hesitation to *Taxites podocarpoides*, is also in the Oxford collection.

Sphenopteris (?) *macrophylla* is the name given by Brongniart to the Stonesfield specimen in the Oxford Museum; he refers the fossil to the ferns with considerable hesitation, and defines the species as follows:—

"Foliis pinnatis, rachi plano (alato ?) pinnulis alternis distantibus, maximis (tripollicaribus) inæquale pinnatifidis, lobis distantibus linearibus obtusis uninerviis."[1]

The specimen thus described is, I believe, specifically identical with those represented on Plate X., and with the fossil figured by Phillips as *Taxites podocarpoides*, Brongn. Brongniart, in his description of this species, adds after the name "(ramuli et fructus)"; he states that it occurs in Jurassic beds, without giving the locality. The specimen figured by Phillips under the French author's name may perhaps be a distinct species. A fragment of a leaf, or leaflet, figured by Buckman in Murchison's *Geology of Cheltenham*, is in all probability an ultimate segment like those represented on Plate X. Saporta refers Brongniart's *Sphenopteris macrophyllus*

---

[1] Brongniart (28^2), p. 212.

to his new genus *Stenopteris*, including the English plant, which he erroneously states[1] was obtained from the Inferior Oolite of Yorkshire, with French specimens in a new species *Stenopteris desmomera*.

The Stonesfield material is too fragmentary to be described as a new species. Some of the examples appear to be identical with the Steierdorf plant named by Ettingshausen *Thinnfeldia speciosa*, a species from a somewhat lower geological horizon than the Stonesfield Slate. The resemblance between Ettingshausen's type-specimen and some of the Stonesfield examples is so close that they may well be specifically identical, but while hesitating to definitely adopt his name, I feel that no useful purpose would be served by referring the English material to a distinct species.

A specimen in the Oxford Museum differs from any in the British Museum in having rather longer and narrower ultimate segments, but it is undoubtedly specifically identical with those figured on Plate X.

We may thus describe the Stonesfield plant:—Fronds bipinnate; strong rachis, giving off linear pinnules which may reach a length of 4·5 cm., but are usually shorter, agreeing in size, shape, and venation with *Thinnfeldia speciosa* as figured by Ettingshausen. The pinnules are traversed by a well-defined midrib from which lateral veins extend obliquely to the edge of the lamina. At the apex of the pinnæ the pinnules are replaced by a terminal lobed lamina with oblique veins (Pl. X. Fig. 1).

V. 4074. Pl. X. Fig. 1.

Imperfect pinna, 11 cm. long, showing the terminal lamina and linear pinnules with a distinct midrib and indications of the lateral veins. The pinnules are decurrent on the stout rachis by their lower margin.

Stonesfield (?).

V. 3422. Pl. X. Fig. 2.

The rachis is represented by a groove with winged borders formed by the decurrent laminæ of the pinnules. Each segment has a distinct midrib and secondary veins. Buckman's *Salicites*

---

[1] Saporta (73), p. 292.

longifolius is probably an imperfect pinnule of a pinna like that shown in Fig. 2. Compare also *Thinnfeldia speciosa*, Ett.

Brodie Coll.

V. 3425. Pl. X. Fig. 3.

The terminal portion of a pinna showing the deep lobing of the lamina as it gradually becomes dissected into linear pinnules.

Eyeford (Stonesfield Slate). *Brodie Coll.*

V. 3424. A specimen similar to that represented in Pl. X. Fig. 1.

Eyeford. *Brodie Coll.*

Other specimens :—**V. 4075, 41383** (Eyeford, *Morris Coll.*).

Group GYMNOSPERMÆ.

Class GINKGOALES.

Genus **GINKGO**, Kaempfer.

[Amœnitates Exoticæ, p. 811, 1712.]

The application of the recent generic name *Ginkgo* to fossil forms has become so thoroughly established, that it seems inadvisable in this case to add the usual termination *ites* when using the designation for extinct types.

Ginkgo digitata (Brongniart).

[Hist. vég. foss. p. 219, pl. lxi. *bis*, figs. 2–3, 1828.]

(Pl. XI. Fig. 3; Text-fig. 9.)

The synonymy of this species is given at length in vol. i. of the *Jurassic Flora* [1]; the following additional records may be added as having reference to leaves which I believe to be identical with *Ginkgo digitata*.

1845. *Noeggerathia* (?), Buckman, Geol. Cheltenham, pl. i. fig. 5.
1863. *Stricklandia acuminata*, Buckman, Geologist, vol. v. p. 395, pl. xx.
1871. *Cyclopteris latifolia*, Phillips, Geol. Oxford, p. 168.
1900. *Ginkgo digitata*, Seward & Gowan, Annals Bot. vol. xiv. p. 140, pl. x. fig. 69.

---

[1] Seward (00), p. 254.

The specimen figured by Buckman as ? *Noeggerathia* is in the British Museum (V. 3433). The lobed and petiolate leaf reproduced from a drawing by Mrs. Strickland in the *Geologist* for 1863 [1] is no doubt specifically identical with those represented in Pl. XI. Fig. 3 and in Text-fig. 9. The original specimen is in

FIG. 9.—*Ginkgo digitata* (Brongn.). Nat. size. From a specimen in the Cirencester Museum.

the Strickland Collection in the Sedgwick Museum, Cambridge. The leaf is well preserved, and shows a deeply bisected lamina with clearly marked veins. The leaf figured in the *Geology of Cheltenham* as *Stricklandia acuminata* may be specifically identical with *Baiera Phillipsi* (Pl. IX. Fig. 2).[2] The leaf figured in the *Geologist*

---

[1] Buckman (63), pl. xx.

[2] See p. 102. Also Seward (00), pl. ix. fig. 4; text-fig. 47.

(original specimen in the Sedgwick Museum), which is also named by Buckman *Stricklandia acuminata*, is, I believe, distinct from that to which the name was originally applied.

The resemblance between the specimens shown in Pl. XI. Fig. 3 and in Text-fig. 9 and some of the leaves from the Inferior Oolite of Yorkshire is so close that we cannot separate them without unduly exaggerating the importance of the slight difference in geological horizon. A comparison may be made between the Cirencester leaf (Text-fig. 9) and the Japanese species described by Yokoyama as *Ginkgodium Nathorsti*;[1] but I do not regard the latter as identical with the type from Yorkshire and Stonesfield.

V. 3429. Pl. XI. Fig. 3.

An imperfect leaf, 5·5 cm. broad, similar to that figured by Buckman in 1845 as ?*Noeggerathia*.[2] The lamina shows a deep median sinus, and is traversed by numerous dichotomously branched veins.

Eyeford (Stonesfield Slate). *Brodie Coll.*

V. 3433. The original of Buckman's ?*Noeggerathia*. A bilobed leaf like that shown in Pl. XI. Fig. 3.

Eyeford. *Brodie Coll.*

Text-fig. 9.

This specimen from Eyeford (Cirencester Museum) is more perfect than the smaller leaf shown in Fig. 3, Pl. XI.; part of the petiole is preserved, and the lamina is broader than in the British Museum specimen. A reduced drawing of this leaf has already been published in the *Annals of Botany* as an example of *Ginkgo digitata*.[3]

Eyeford. *Cirencester Museum Coll.*

Other specimens from Eyeford:—**V. 3427** (veins fairly distinct), **V. 3430, V. 3434.**

Brodie Coll.

---

[1] Yokoyama (89), pl. viii.

[2] Murchison (45), pl. i. fig. 5.

[3] Seward & Gowan (00), pl. x. fig. 69.

Genus **BAIERA**, Braun.

[Münster's Beiträge, Heft vi. p. 20, 1843.]

Baiera Phillipsi, Nathorst.

[Öfvers. k. Veten. Akad. Förh. p. 76, 1880.]

(Pl. IX. Fig. 2; Text-fig. 10.)

1829. *Sphenopteris longifolia*, Phillips, Geol. Yorks. p. 148, pl. vii. fig. 17.
1845. *Stricklandia acuminata*, Buckman, Geol. Cheltenham, pl. ii. fig. 2.
1854. *S. acuminata*, Morris, Brit. Foss. p. 23.
1864. *Baiera longifolia*, Leckenby, Quart. Journ. Geol. Soc. vol. xx. p. 76.
1871. *Stricklandia acuminata*, Phillips, Geol. Oxford, p. 169.
1875. *Cyclopteris longifolia*, Phillips, Geol. Yorks. p. 200, pl. vii. fig. 17.
1880. *Baiera Phillipsi*, Nathorst, Öfvers. k. Vet. Akad. Förh. p. 76.
1892. *B. longifolia*, Fox-Strangways, Tab. Foss. p. 137.
1894. *Stricklandia acuminata*, Woodward, Lower Ool. p. 598.
1900. *Baiera Phillipsi*, Seward, Jurass. Flor. i. p. 269, pl. ix. fig. 4; text-fig. 47.

A specimen in the British Museum collection (V. 3433; Pl. IX. Fig. 2) is probably the original of the drawing given in the *Geology of Cheltenham* under the name *Stricklandia acuminata*.

The two specimens represented in Text-fig. 10 and in Fig. 2, Pl. IX. are the only examples from the Stonesfield Slate that I have seen which can be referred to the genus *Baiera*. Without more material it is hardly possible to determine the species with confidence, but the imperfect leaves agree so closely with the Inferior Oolite specimens for which Nathorst's name has been adopted, that they may be reasonably referred to his species. The fossil shown in Text-fig. 10 might be taken for *Sagenopteris*, but the lobed character of the lamina and its probable identity with the more complete example shown in Pl. IX. Fig. 2 lead me to speak of it without hesitation as *Baiera*.

The drawing of *Stricklandia acuminata* in the *Geology of Cheltenham* would be valueless as an aid to identification were it not probable that the specimen (V. 3433) reproduced on Pl. IX. Fig. 2 represents the original leaf.

In an article by an anonymous writer (Buckman) in the *Geologist* of 1863, a drawing by Mrs. Strickland is given of a leaf from Sevenhampton, said to be identical with *Stricklandia acuminata*;[1] but this specimen (now in the Sedgwick Museum, Cambridge) has broader lobes and is probably specifically identical with the leaf shown in Pl. XI. Fig. 3, which I have named *Ginkgo digitata.*

V. 3433. Pl. IX. Fig. 2.

Probably the original of Buckman's figure of *Stricklandia acuminata* (*Geol. Cheltenham*, pl. ii. fig. 2). An obscure impression on sandstone, showing part of a petiole and two spreading linear lobes of a lamina traversed by several indistinct veins. *Brodie Coll.*

Fig. 10.—*Baiera Phillipsi*, Nath. V. 3422. Nat. size.

V. 3422. Text-fig. 10.

An obscure impression similar to that represented in Pl. IX. Fig. 2, but showing more segments. Compare fig. 4, pl. ix., *Jurassic Flora*, vol. i.; there can be little doubt as to the identity of the Yorkshire Coast species, *Baiera Phillipsi*, Nath., with the imperfect specimen from Stonesfield. *Brodie Coll.*

[1] Buckman (63), pl. xx.

Class CYCADOPHYTA.[1]

Section BENNETTITALES.[2]

In the second volume of the *Wealden Flora* (British Museum Catalogue) an account was given of the history of our knowledge of the genera *Williamsonia* and *Bennettites*,[3] and I expressed the opinion that the fossils described by Williamson under Carruthers' designation *Williamsonia* are in all probability generically identical with the reproductive shoots of *Bennettites*, as represented by *B. Gibsonianus*, Carr. It is not proposed to traverse the same ground again, but to draw attention to some more recent opinions bearing on the nature of these interesting types of extinct Cycadophyta.

There has been considerable difference of opinion as to the correctness of Williamson's conclusions which he expressed in a restoration of *Williamsonia*;[4] he brought forward evidence in support of the view that the pinnate Cycadean fronds described by Lindley & Hutton as *Zamia gigas* belong to the plants which bore the reproductive shoots known as *Williamsonia*. Saporta believed that the leaves and reproductive organs had no connection with one another, and Solms-Laubach demurred to the acceptance of Williamson's views without further evidence. In a paper communicated to the Philosophical Society of Cambridge in 1897[5] I described some Yorkshire specimens of *Zamites gigas* and *Williamsonia* included in the Yates Collection in the Natural History Museum, Paris, and expressed my conviction that these specimens demonstrate the correctness of the statement that *Williamsonia* and *Zamites gigas* are the female inflorescences and fronds of one and the same plant. In the volume on the *Jurassic Flora* published in 1900 I again discussed the connection between *Williamsonia* and *Bennettites*, and spoke of them as two very

---

[1] Nathorst (02), p. 3.
[2] Engler & Prantl (97), pp. 5, 341.
[3] Seward (95), p. 134.
[4] Williamson (70).
[5] Seward (97^2).

closely allied, if not identical genera. It may, as I have already suggested, be a convenience to retain both designations, or to use *Williamsonia* as a subgenus of *Bennettites*. Nathorst has also expressed the opinion that the two types may be closely allied. Another question which has been dealt with by several authors is the precise nature of the Williamsonian inflorescence: Williamson believed that the specimens of *Williamsonia* which he described represented both male and female flowers, and that the plant *Williamsonia gigas* agreed with existing Cycads in being diœcious. Solms-Laubach expressed the opinion that Williamson's specimens represent the male flowers, and this view was taken also by Saporta and Nathorst. On the other hand, I stated my opinion that the Yorkshire fossils described as *Williamsonia gigas* are imperfectly preserved female inflorescences. This conclusion has been arrived at also by Professor Lignier,[1] whose work has done much to increase our knowledge of the Bennettitales; he has recently published a memoir based, in part, on an examination of the Yates specimens in Paris, in which he states his opinion in favour of regarding *Williamsonia gigas*, as represented by the reproductive shoots described by Williamson, as the female inflorescence. The fossils referred to by Williamson as male flowers Lignier believes to represent portions of ovuliferous flowers. During the last few years Mr. Wieland,[2] of Newhaven, has published several papers on the reproductive organs of American plants belonging to the genus *Bennettites*, and has contributed most important facts in regard to the nature of the reproductive organs. He has demonstrated the occurrence of fertile shoots bearing in the centre more or less aborted ovules, and at the periphery pinnate leaves with numerous sporangia agreeing closely with the sori of recent Marattiaceous ferns. I hope to deal at greater length with the general morphology and systematic position of the Bennettitales in a memoir, now in preparation, to be published by the Palæontographical Society.

The Jurassic rocks of England dealt with in this volume have not afforded any specimens of Bennettitean flowers, with the

---

[1] Lignier (03).
[2] Wieland (99), (99^2).

exception of the fossil originally described by Buckland as *Podocarya.*[1] Buckland's type-specimen cannot be found, but there is practically no doubt that it must be assigned either to the genus *Williamsonia* or to *Bennettites.*

Genus **WILLIAMSONIA**, Carruthers.

[Trans. Linn. Soc. vol. xxvi. p. 680, 1870.]

Williamsonia Bucklandi (Unger).

[Gen. spec. foss. plant. p. 327, 1850.]

1837. *Podocarya*, Buckland, Geol. and Min. vol. i. p. 503; vol. ii. p. 101, pl. lxiii. figs. 2–10.
1849. *Podocarya*, Brongniart, Tableau, p. 106.
1850. *P. Bucklandi*, Unger, Gen. spec. foss. plant. p. 327.
1872. *P. Bucklandi*, Schimper, Trait. pal. vég. vol. ii. p. 477.
1873. *P. Bucklandi*, Zigno, Flor. foss. Oolit. vol. ii. p. 5.
1878. *P. Bucklandi*, Lyell, Elements Geol., 2nd ed., p. 335, fig. 350.
1891. *Williamsonia Bucklandi*, Saporta, Pal. Franç. vol. iv. p. 127, pl. ccxxxviii. figs. 1–3; pl. ccxxxix. fig. 1.
1894. *Podocarya Bucklandi*, Woodward, Lower Ool. p. 598.
1895. *Williamsonia Bucklandi*, Seward, Wealden Flora, ii. pp. 147, 151, 154.
1902. *W. Bucklandi*, Nathorst, K. Svensk. Vet.-Akad. Hand. vol. xxxvi. No. 4, p. 15.
1903. *W. Bucklandi*, Lignier, Mém. Soc. Linn. Normandie, vol. xxi. pp. 22, 29.

It is unfortunate that Buckland's type-specimen is not forthcoming. I made an unsuccessful search among the Jurassic fossils in the Oxford Museum, and Professor Sollas, who kindly undertook to make every effort to find the fossil, was also unable to trace Buckland's specimen. Buckland's specimen was obtained from rocks of Inferior Oolite age at Charmouth, in Dorsetshire. He describes it as an "unique and beautiful fruit," which he regarded as closely allied to the recent Pandanaceæ. Its size is described as being that of a large orange. It consists of a thick central axis or receptacle extending rather more than half-way through the middle of the 'fruit'; from the axis are given off crowded and slender stalks, some of which appear to be terminated by oval

[1] Buckland (37).

seeds. In surface-view the fossil presents the appearance of a regular network of ridges agreeing exactly with the surface-features of *Williamsonia* and *Bennettites*. There can be no doubt that *Williamsonia Bucklandi* is a female inflorescence of the Bennettitean type; it was most probably borne at the apex of a lateral shoot, and consisted of numerous sterile appendages and ovuliferous peduncles attached to a stout central receptacle, the meshes seen on the surface being formed by the contiguous distal ends of the sterile appendages or scales. Sowerby's drawings, published by Buckland, show several of the seeds exposed on the somewhat worn outer surface, and it is clear that the relation between the seeds and interseminal scales is precisely that which has been described in other examples of the Bennettitales. A comparison of Buckland's figures with those given by Carruthers, Solms-Laubach, Lignier, Wieland, and myself demonstrates the Bennettitean nature of the Charmouth specimen.

Williamsonia pecten (Phillips).

[Geol. Yorks. p. 148, pl. vii. fig. 22, 1829.]

(Pl. IX. Fig. 6; Pl. XII. Fig. 8.)

1823. *Fucoides (Caulerpa) pennatula*, Brongniart, Mém. soc. d'hist. nat. Paris, i. p. 301, pl. xxi. fig. 3.
Polypodiolites pectiniformis, Sternberg, Flor. Vorwelt, iii. p. 36, pl. xxxiii. fig. 1.
1825. *Fucoides pennatula*, Sternberg, op. cit. iv. p. vi.
1828. *F. pennatulus*, Brongniart, Hist. vég. foss. p. 49.
Zamia pectinata, Brongniart, Prodrome, pp. 94, 199.
1835. *Z. pectinata*, Lindley & Hutton, Foss. Flor. pl. clxxii.
Filicites scolopendrioides, op. cit. pl. ccxxix.
1836. *Zamia taxina*, op. cit. pl. clxxv.
1838. *Cycadites plumula*, Sternberg, Flor. Vorwelt, vii. p. 195.
1841. *Ptilophyllum pectinatum*, Morris, Ann. Mag. vol. vii. p. 117.
P. taxinum, op. cit. p. 118.
1842. *Encephalartos taxinus*, Miquel, Mon. Cycad. p. 61.
E. pectinatus, ibid.
1843. *Pterophyllum Preslianum*, Göppert, Foss. Cycad. p. 51.
P. taxinum, ibid.
1845. *P. taxinum*, Unger, Synopsis, p. 156.
1847. *P. Preslianum*, Unger, Chlor. Prot. p. lxiii.
1848. *Palæozamia pectinata*, Bronn, Ind. Pal. p. 397.
P. taxina, ibid.

1849. *Zamites taxina*, Brongniart, Tableau, p. 106.
Z. pectinata, ibid.
1850. *Pterophyllum Preslianum*, Unger, Gen. spec. p. 288.
P. taxinum, op. cit. p. 289.
1851. *Dioonites plumula*, Miquel, Over Rangs. foss. Cycad. p. 212.
D. taxinus, ibid.
1854. *Palæozamia pectinata*, Morris, Brit. Foss. p. 15.
P. taxina, ibid.
1856. *Dioonites plumula*, Bornemann, Foss. Cycad. p. 56.
D. taxinus, ibid.
1871. *Palæozamia pectinata*, Phillips, Oxford, p. 169, Diag. xxx. figs. 2–3.
P. taxina, op. cit. p. 169, Diag. xxx. figs. 4–5.
1872. *P. pectinata*, Balfour, Pal. Bot. p. 80, fig. 79.
1873. *P. pecten*, Sharp, Quart. Journ. Geol. Soc. vol. xxix. p. 295.
Zamites pectinatus, Zigno, Flor. foss. Oolit. vol. ii. p. 37.
Pterophyllum taxinum, op. cit. p. 17.
1875. *Palæozamia pectinata*, Judd, Geol. Rutland, p. 276.
1884. *Ptilophyllum acutifolium*, var. *β*, *pectinatum*, Richards, Synopsis, pp. 4–5.
P. acutifolium, var. *taxinum*, ibid. p. 5.
1894. *P. acutifolium*, Woodward, Lower Ool. p. 599.
Williamsonia pecten, ibid.
1900. *W. pecten*, Seward, Manchester Lit. Phil. Soc. vol. xliv. p. 20, pl. iii. fig. 6.
W. pecten, Seward, Jurassic Flora, i. p. 190, pl. ii. fig. 7; pl. iii. text-figs. 30–35.

The original specimen of *Zamia pectinata* of Lindley & Hutton, from Stonesfield, is in the Oxford Museum; the specimen figured by Phillips as *Palæozamia pectinata*, also from Stonesfield, is in the same collection.

Localities and Horizons.—Stamford (Inferior Oolite); Wansford [Inferior Oolite (Lincolnshire Limestone)]; Stonesfield [Great Oolite (Stonesfield Slate)].

For a diagnosis and general account of the fronds of *Williamsonia pecten* reference must be made to the *Jurassic Flora*, vol. i.[1]

The specimen from Stamford shown in Pl. XII. Fig. 8 undoubtedly belongs to this species; it agrees exactly with the smaller fronds from the Yorkshire plant-beds in which *Williamsonia pecten* is exceedingly common, and is represented by fronds which vary considerably in size and in the shape of the segments. Lindley & Hutton refer the Stonesfield frond to *Zamia pectinata*

[1] Seward (00), p. 190.

without speaking definitely as to its systematic position; the specimen of which a drawing is given in the *Fossil Flora* (pl. clxxii.) is the same which Sternberg described as *Polypodiolites pectiniformis*.

The original of Phillips' figure (Diag. xxx. fig. 2, *Geol. Oxford*) is a larger and more perfect specimen than the drawing suggests; it has been figured also by Sternberg.[1]

In 1823 Brongniart described a Stonesfield fossil as *Fucoides* (*Caulerpa*) *pennatula*;[2] in his *Histoire* (1828)[3] he says that his original determination was based on a drawing sent to him by Buckland, and adds that an examination of the specimens in the Oxford Museum convinced him that they were Cycadean fronds identical with the example figured by Sternberg as *Polypodiolites pectiniformis*. The Oxford Collection includes several examples of *Williamsonia pecten* from Stonesfield in addition to those already referred to; one of them is a portion of a frond 22·5 cm. long and about 4 cm. broad, with narrow and contiguous pinnæ bearing a very close resemblance to Feistmantel's Indian fronds which it has been customary to include in a distinct genus, *Ptilophyllum*. I have elsewhere given reasons for uniting some of the Indian species of *Ptilophyllum* with *Williamsonia pecten*.[4]

Specimens of *Williamsonia pecten* from Stonesfield are represented also in the Sedgwick Museum, Cambridge, and in other collections.

V. 3361. Pl. IX. Fig. 6.

A piece of a frond 9 cm. long and 2·5 cm. broad. The axis is represented by a deep groove. The pinnæ are crowded and slightly imbricate, with blunt apices, and traversed by 8–10 parallel veins.

Stonesfield. *Brodie Coll.*

52,868. Pl. XII. Fig. 8.

A portion of a narrow and possibly a young frond, very similar to specimens from the Yorkshire coast.[5]

Stamford (Inferior Oolite). *Sharp Coll.*

---

[1] Sternberg (23), pl. xxxiii. fig. 1.
[2] Brongniart (23), p. 313, pl. xxi. fig. 3.
[3] Brongniart (28), p. 49.
[4] Seward (00), p. 192; (03^2).
[5] Seward (00), pl. iii. fig. 1.

V. 84. Portion of a narrow frond.
Stonesfield. *Egerton Coll.*

V. 3521. Similar to V. 3361 (Pl. IX. Fig. 6); but the pinnæ are smaller and have fewer veins.

Other specimens:—**V. 4659, V. 4664, V. 4668, V. 4670** (from the Inferior Oolite of Wansford, Northants), **V. 9708, 40,693, 40,694, 41,377.**

Genus CYCADEOIDEA, Buckland.

[Proc. Geol. Soc. London, vol. i. No. 8, pp. 80–81, 1827.]

In accordance with the plan adopted in the description of a Cycadean stem from the Lower Lias of Dorsetshire, I propose to substitute the generic name *Cycadeoidea* for *Bucklandia* in describing the Stonesfield stem originally figured by Sternberg as *Conites Bucklandi*.

Cycadeoidea squamosa (Brongniart).

[Prodrome, p. 128, 1828.]

1823. *Conites Bucklandi*, Sternberg, Flor. Vorwelt, iii. p. 36, pl. xxx.
1825. *C. Bucklandi*, op. cit. iv. p. xxxix.
1828. "Amentum of a *Cycadeoidea*," Buckland, Trans. Geol. Soc. [ii.], vol. ii. p. 400.
Bucklandia squamosa, Brongniart, Prodrome, p. 128.
1843. *Cycadites Bucklandi*, Göppert, Ueber foss. Cycad. p. 37.
1845. *Bucklandia squamosa*, Unger, Syn. p. 169.
1849. *Cycadeoidea squamosa*, Brongniart, Tableau, p. 106.
1850. *Bucklandia squamosa*, Unger, Gen. spec. foss. plant. p. 315.
1854. *B. squamosa*, Morris, Brit. Foss. p. 3.
1870. *B. squamosa*, Carruthers, Trans. Linn. Soc. vol. xxvi. p. 686.
Clathraria Bucklandi, Schimper, Trait. pal. vég. vol. ii. p. 183.
1871. *Bucklandia squamosa*, Phillips, Geol. Yorks. p. 170, Diag. xxix.
1885. *B. squamosa*, Zigno, Flor. foss. Oolit. vol. ii. p. 182.
1894. *Clathraria Bucklandi*, Woodward, Lower Ool. p. 599.

The type-specimen from Stonesfield, as figured by Sternberg and by Phillips, is in the Oxford Museum.

In the diagnosis given by Sternberg the fossil is regarded as a cone, and described as follows:—

"C. strobilo ovato 8 pollices longo, 4 poll. crasso, squamis extus convexis, interna concavis apice subtruncatis vetusisve, axi pollicum fere crassa, squamulis minutis tecta."

In 1828 Buckland refers to the specimen, figured by Sternberg from Sowerby's drawing, as an amentum of a *Cycadeoidea*. Brongniart, in his *Prodrome*, includes the Stonesfield stem in his genus *Bucklandia*. Carruthers compares the stem with those of recent Cycads, and adds that the data afforded by the single specimen are insufficient to enable him to refer it with certainty to its proper genus.[1]

The type-specimen is in the form of a cast 18 cm. long; in the lower part it shows a pith-cast 2–3 cm. broad, with irregular lozenge-shaped projections which represent the meshes of the inner face of the cylinder of wood. Several thick, crowded, and imbricate leaf-bases are well preserved, and exhibit a striking similarity to the persistent petiole-bases on the stems of *Encephalartos* and other existing Cycads. There is no trace of any lateral fertile axis, as in the typical *Bennettites*. A pith-cast very like that of *Cycadeoidea squamosa* has been figured by Lignier from the Lias of Ste. Honorine-la-Guillaume as *Cycadeomyelon densecristatum*.[2]

V. 4669. A portion of a medullary cast of a Cycadean stem, possibly *Cycadeoidea squamosa*.

Stonesfield. *Egerton Coll.*

Genus **ZAMITES**, Brongniart.

[Prodrome, p. 94, 1828.]

For a discussion on the use of this generic term reference may be made to vol. ii. of the *Wealden Flora*.[3]

Zamites megaphyllus (Phillips).

[Geol. Oxford, p. 169, Diag. xxx. fig. 1, 1871.]

(Pl. X. Figs. 4–5; Pl. XII. Figs. 1, 3–5; Text-fig. 11.)

1871. *Palæozamia megaphylla*, Phillips, Geol. Oxford, p. 169, Diag. xxx. fig. 6.

P. longifolia, op. cit. p. 169, Diag. xxx. fig. 6.

1873. ? *Yuccites Schimperianus*, Zigno, Flor. foss. Oolit. vol. ii. p. 7, pl. xxvi.

---

[1] Carruthers (70), p. 687.

[2] Lignier (95), pl. vii. fig. 7.

[3] Seward (95), p. 75.

1880. *Yuccites*, sp., Nathorst, Öfvers. k. Vet.-Akad. Förh. 1880, p. 79.
1884. *Yuccites*, sp., Richards, Synopsis, p. 8.
Podozamites lanceolatus, var. *geminus*, op. cit. p. 3.
1885. ? *Krannera mirabilis*, Velenovsky, Gymnosp. Böhm. Kreid. pls. i.–iii.
1894. *Yuccites megaphylla*, Woodward, Lower Ool. p. 598.

The type-specimens of Phillips' *Palæozamia megaphylla* and *P. longifolia* are in the Oxford Museum; the former is redrawn in Pl. XII. Fig. 4, and the latter in Text-fig. 11.

Frond pinnate; pinnæ, attached to the rachis at a wide angle, reaching a length of more than 30 cm. and a breadth of over 3 cm., linear in form, attached by a slightly contracted concave basal surface, which is somewhat thickened; the lamina tapers gradually to an elongated acuminate apex, frequently falcate near the tip, one margin being practically straight, while the other is curved (Pl. XII. Figs. 4–5). Veins numerous and parallel, converging slightly towards the base and apex of the pinna.

The material on which this diagnosis is based consists almost entirely of detached pinnæ, and it is impossible therefore to give a complete account of the form of the frond. The specimen figured by Phillips as *Palæozamia longifolia* represents a portion of a pinnate frond; it is more accurately reproduced in Text-fig. 11. It is impossible to speak with certainty as to the connection between this pinnate fragment and the detached leaflets of the type named by Phillips *Palæozamia megaphylla* (Pl. XII. Fig. 4). The former specimen bears a close resemblance to some species of *Podozamites*, e.g. *P. lanceolatus*, but the examination of a considerable number of detached pinnæ, varying in size from 3·5 to over 30 cm. in length, leads me to regard *Palæozamia megaphylla* and *P. longifolia* as specifically identical. The great length of the pinnæ, like that of which a portion is shown in Fig. 5, and their shape can be matched almost perfectly with pinnæ of recent Cycadean fronds, e.g. *Ceratozamia mexicana*. On the whole I incline to the opinion that the numerous Monocotyledon-like leaves from Stonesfield are the pinnæ of a Cycadean frond. In certain recent Cycads, e g. species of *Encephalartos*, the pinnæ may be readily detached from the rachis by a well-defined separation-surface, and the narrow oval scars left on the axis of the frond are such as might be formed by the bases of pinnæ like that represented in Fig. 4, Pl. XII.

Both the British Museum and the Oxford Museum collections include several examples of the pinnæ of *Zamites*. In many of them the base is well shown: it has the form of a clean-cut margin terminating the pinna by a curved line, which may extend sufficiently far into the lamina to present the appearance of a semicircle. In some of the larger pinnæ the base is slightly over 2 cm. in breadth, and is thus greater than that of the pinnæ of recent Cycads. Another characteristic feature is the very gradually tapered apical portion of the pinna, and, as seen in the drawing (Pl. XII. Fig. 5), the surface of the lamina is strongly convex in its narrow part: this convexity is no doubt the result of contraction of the stout leathery pinna after death. The occurrence of pinnæ which exhibit identical features, varying in length from 3·5 to 33 cm., is an additional piece of evidence in favour of regarding them as segments of large Cycadean fronds. The parallel veins are clearly shown in the fragments of lamina represented in Pl. XII. Fig. 1, and in Pl. X. Fig. 4.

The long parallel-veined leaves figured by Zigno from Jurassic rocks of Italy as *Yuccites Schimperianus* are probably identical with the Stonesfield Slate pinnæ. Other fossils comparable with *Zamites megaphyllus* are *Zamia washingtoniana*,[1] a Potomac species described by Ward, leaves from the Lower Cretaceous of Bohemia referred by Velenovsky to the genus *Krannera*[2] (*K. mirabilis*, Cord.), and Schenk's species *Pterophyllum giganteum*,[3] from the Trias Raibl beds. One may note also a resemblance between the Stonesfield pinnæ and the long narrow leaves from the Rhætic of Tonkin identified by Zeiller as *Noeggerathiopsis Hislopi*.[4]

Nathorst has suggested that Phillips' species *Palæozamia megaphylla* (= *Zamites megaphyllus*) should be referred to *Yuccites*.[5]

V. 86. Pl. X. Fig. 4.

A small piece of a broad pinna 16 cm. long and 4 cm. broad. The lamina is traversed by numerous parallel veins between which,

---

1 Ward (95), pl. ii. fig. 6.
2 Velenovsky (85), pls. i.–iii.
3 Schenk (66), pl. ii.
4 Zeiller (02), pl. xl.
5 Nathorst (80), p. 79.

as shown in the drawing, occur finer longitudinal lines, which probably indicate strands of mechanical tissue between the true veins. The specimen is incomplete; at one end the lamina is 4 cm. broad, and at the other 3·6 cm.

Stonesfield. *Egerton Coll.*

V. 4064. Pl. X. Fig. 5.

Part of a leaf 13·5 cm. long, and 2·7 cm. broad at the widest end. The parallel veins and the fainter intercostal lines are less clearly shown than in the specimen represented in Fig. 4. One edge (right) of the pinna is practically straight, while the other is curved, a character frequently noticed in the segments of Cycadean fronds. The circular scar near the broader end of the fossil is simply a hole in the lamina.

Stonesfield Slate.

V. 4067. Pl. XII. Fig. 5.

A pinna 22 cm. long and 3·1 cm. broad at the widest end, but tapering to a breadth of 4 mm. at the apical portion. Veins as in the specimen represented in Fig. 4, Pl. IX. This specimen, like many others, shows that the lamina, which is almost or quite flat at the broader end, becomes gradually more convex towards the narrower end. This strongly arched form of the apical portion is no doubt the result of contraction of the dead pinna.

Stonesfield.

V. 4652. Pl. XII. Fig. 3.

Possibly a complete pinna, 8·5 cm. long; at the broader end the lamina is somewhat contracted, and presents an appearance which suggests that it is the actual base.

V. 3460. Pl. XII. Fig. 1.

The broad truncate end of a pinna, showing the parallel veins extending to the edge of the lamina. On the same piece of rock occurs another fragment which represents an acuminate apex of a lamina.

Sevenhampton. *S. Buckman Coll.*

Pl. XII. Fig. 4.

A specimen in the Oxford Museum, and probably the type of Phillips' *Palæozamia megaphylla*. The pinna, which is 13 cm. long, shows a slight callosity at the base and the tapered convex distal portion of the lamina.

Stonesfield.

Text-fig. 11.

A specimen in the Oxford Museum; the original of Phillips' *Palæozamia longifolia*. As pointed out above, this example may possibly be a fragment of *Podozamites lanceolatus* (L. & H.), but on the whole I prefer to regard it as a piece of a frond of *Zamites megaphyllus*.

Stonesfield.

Fig. 11.—*Zamites megaphyllus* (Phill.). Nat. size. From a specimen in the Oxford Museum.

V. 4066. An incomplete pinna 21 cm. long, 3 cm. broad at the wider end, tapering to 1·3 cm. at the narrower end. The narrower part of the lamina is much more convex in section than the broader end.

V. 9010. A specimen 18 cm. long and 5 cm. broad; a portion only of the pinna is seen; venation clearly shown, here and there forked veins may be recognised.
Stonesfield ?

Other specimens:—**V. 3426, V. 3428, V. 3460, V. 3461, V. 3462, V. 4063, V. 4065** (the apical portion of a pinna, labelled "Termination of the rib of a Saurian"), **V. 4068, V. 4644, V. 4650, V. 4651, V. 4654** (?), **V. 4675, V. 6583** (?), **41,380, 52,817.**

Genus **CTENIS**, Lindley & Hutton.

[Foss. Flor. vol. ii. pl. ciii. 1834.]

Ctenis latifolia (Brongniart).

[Hist. vég. foss. p. 266, pl. lxxxii. fig. 6.]

1828. *Tæniopteris latifolia*, Brongniart, Prodrome, p. 62.
T. latifolia, Brongniart, Hist. vég. foss. p. 266, pl. lxxxii. fig. 6.
1833. *Odontopteris latifolia*, Sternberg, Flor. Vorwelt, v.–vi. p. 79.
1838. *Tæniopteris latifolia*, ibid. vii. p. 139.
Zamites latifolius, ibid. p. 199.
1845. *Tæniopteris latifolia*, Unger, Synopsis, p. 37.
Cycadites (?), Buckman, Geol. Cheltenham, p. 67, pl. i. fig. 2.
1848. *Tæniopteris latifolia*, Bronn, Ind. Pal. p. 1214.
1849. *T. latifolia*, Brongniart, Tableau, p. 105.
1850. *T. latifolia*, Unger, Gen. spec. foss. plant. p. 213.
1854. *T. latifolia*, Morris, Brit. Foss. p. 23.
1856. *T. latifolia*, Zigno, Flor. foss. Oolit. vol. i. p. 204.
1869. *Oleandridium vittatum*, Schimper, Trait. pal. vég. vol. i. p. 608.
1871. *Tæniopteris latifolia*, Phillips, Geol. Oxford, p. 168.
Pterophyllum Buckmanni, ibid. p. 170.
1873. *Tæniopteris vittata* (pars), Saporta, Pal. Franç. vol. i. p. 445.
1884. *Pterophyllum Buckmanni*, Richards, Synopsis, p. 2.
1894. *P. Buckmanni*, Woodward, Lower Ool. p. 599.
Tæniopteris vittata, ibid. p. 600.

The type-specimen of Brongniart's *Tæniopteris latifolia* is in the Oxford Museum. The veins are incorrectly represented in the figure as separate; they agree precisely with the veins of *Ctenis* in being joined together laterally by oblique cross-connections. The single example of *Ctenis* that I have seen from Stonesfield is too imperfect to form the basis of a satisfactory diagnosis. The type-specimen consists of a torn and incomplete piece of a frond, with a broad rachis giving off short and broad segments traversed by parallel veins, which occasionally fork and are connected here and there by oblique branches. The veins are, on the average, rather more than 1 mm. apart.

I have no doubt as to the specific identity of the specimen figured by Buckman from Sevenhampton as *Cycadites*? with the type-specimen of Brongniart.

Ctenis latifolia differs from the Inferior Oolite species *C. falcata*[1] in its shorter and broader segments; as regards venation-characters the two forms are identical. An imperfect leaflet, which I described as *Ctenis*, sp., from a specimen in the Manchester Museum,[2] may be compared with the Stonesfield type, but we have not sufficient data from which to decide as to their possible identity. A form of *Ctenis* frond similar to *C. latifolia* has been described by Raciborski from the Lower Jurassic rocks of Cracow as *Ctenis Zeyschneri*,[3] which is no doubt a closely allied plant; another similar type is represented by *Ctenis orovallensis*,[4] a species figured by Lester Ward from the Jurassic beds of Oroville, California. It is by no means unlikely that the Stonesfield plant is specifically identical with the American form, but until better specimens are obtained it is impossible to make a more definite statement as to the position and characters of the Stonesfield type. Phillips includes *Pterophyllum comptum*, Phill. [= *Nilssonia compta* (Phill.)], in his list of Stonesfield species; it is possible that he applied this name to *Ctenis latifolia*.

Genus **OTOZAMITES**, Braun.

[Münster, Beit. Petrefact. Heft vi. p. 36, 1843.]

Otozamites, sp.

[Cf. *O. obtusus* (L. & H.), var. *ooliticus*.]

The Museum collection includes a few imperfect and obscure specimens of *Otozamites* fronds, which are too small and indistinct

---

1 Seward (00), pl. viii. fig. 2.
2 Seward (00), p. 232, fig. 42; also (00[2]), pl. ii.
3 Raciborski (94), pls. xvi.–xvii.
4 Ward (00), p. 357, pl. lviii. fig. 4.

to determine with confidence. These fragments, which were obtained from the Inferior Oolite of Barnack and Wansford, in Northamptonshire, are perhaps best described as *Otozamites*, sp., and may be identical with the Yorkshire type of frond described as *O. obtusus*, var. *ooliticus*.[1]

The Jermyn Street Museum contains a specimen of *Otozamites* from the Inferior Oolite of Stamford (a locality on the borders of Lincolnshire and close to Barnack and Wansford), bearing the name *Otozamites graphicus*, a species recorded by Professor Judd[2] from the Lincolnshire Oolite of Stibbington. The Stamford specimen may be specifically identical with those in the British Museum, but it is impossible in the absence of better specimens to determine its exact position.

V. 4641, V. 4660. Pieces of *Otozamites* fronds.

Sheep pit near Wansford (Inferior Oolite). *C. W. Peach Coll.*

V. 6585. An obscure impression, 17 cm. long; pinnæ contiguous, with auriculate upper basal edge.

Stonesfield Slate.

51,129. A similar specimen from Barnack. *Morris Coll.*

V. 5274, V. 6588. Obscure impressions; locality unknown.

Genus **SPHENOZAMITES**, Brongniart.

[Tableau, p. 61, 1849.]

Brongniart proposed the name *Sphenozamites* as a sectional designation for Cycadean fronds included under *Otozamites* and characterised by the presence of divergent veins and by the absence of an auriculate base; he suggested that the name might eventually be raised to generic rank. Zigno made use of Brongniart's genus in his *Oolitic Flora*, and instituted the species *Sphenozamites Rossii* for a pinnate frond with broad wedge-shaped pinnæ traversed by spreading veins, and without the auriculate

---

[1] Seward (00), p. 218, pl. i. fig. 1; pl. ii. fig. 2.
[2] Judd (75), p. 174.

base that forms a generic character of the typical *Otozamites*. Schimper also adopts *Sphenozamites* as a generic name, but correctly points out that *Otozamites Beani* of Lindley & Hutton, which Brongniart gave as the type of *Sphenozamites*, is a true *Otozamites*. Saporta defines the genus as follows:—

"Frondes plerumque rachi valida tereti instructæ pinnatæ, pinnæ vel foliola latiores majusculæ basi plus minusve angustata subpedicellatæ æquilaterales cartilagineocinctæ integræ aut apice sinuatæ dentatæque spinosæ, racheos lateribus ordine alterno insertæ nec unquam basibus inter se connexis racheos superficiem tegentes, nervulis e loco insertionis radiantibus numerosis dichotome pluries divisis."[1]

Among recent Cycads we find a similar form of pinna in species of *Encephalartos* and *Zamia*, e.g., *Z. Skinneri*, *Z. integrifolia*, *Z. muricata*.

Fronds conforming to the *Sphenozamites* pattern occur in Lower Permian rocks,[2] and extend through the Jurassic system. It is a convenience to adopt this generic name for fronds bearing pinnæ having a comparatively broad and short, more or less wedge-shaped lamina; the distinction between certain forms of *Otozamites* and *Sphenozamites* is often very slight, and probably does not constitute a generic difference in the sense in which genera are defined among existing plants. *Sphenozamites*, like *Otozamites* and numerous other genera, is used in a provisional and artificial sense as denoting certain characteristic features in the shape of the leaflets, and not in the sense of generic designations based on characters of primary taxonomic importance. In the *Wealden Flora*, vol. ii., I described a plant under the name of *Withamia*, which possesses broad pinnæ or leaves very similar to those of *Sphenozamites*, but distinguished by the presence of strong recurved hooks situated immediately below each leaf-like appendage. Professor Zeiller[3] afterwards substituted the generic name *Sewardia* for *Withamia*, as I had overlooked the fact that the latter term had previously been used by Unger.

---

[1] Saporta (75), p. 181.
[2] Zeiller (00), p. 232; Renault (81).
[3] Zeiller (00).

Sphenozamites Belli, sp. nov.

(Pl. XI. Fig. 4; Text-fig. 12.)

1845. ? *Naiadea obtusa*, Buckman, Geol. Cheltenham, p. 67, pl. i. fig. 2.
1854. *N. obtusa*, Morris, Brit. Foss. p. 12.
1871. *N. obtusa*, Phillips, Geol. Oxford, p. 169.
1894. *N. obtusa*, Woodward, Lower Ool. p. 598.

Type-specimen (Text-fig. 12) formerly in the collection of A. M. Bell, Esq., and recently presented by him to the Oxford Museum. From Stonesfield.

The species is founded on detached pinnæ characterised by their wedge-shaped lamina, which tapers gradually to a comparatively acute apex, one edge of the lamina being more strongly arched than the other; the pinnæ are contracted at the proximal end, and were probably attached by a narrow base to the rachis. Veins numerous, dichotomously branched, spreading from the base through the substance of the lamina.

The leaflet figured by Buckman as *Naiadea obtusa* is probably specifically identical with Mr. Bell's specimen, but as I have not been able to identify the original of the crude drawing published in the *Geology of Cheltenham* it seemed wiser to institute a new name. I have therefore named the species after Mr. Bell, whose collection of fossils has afforded me much assistance in the investigation of the Stonesfield flora. *Sphenozamites Belli* may be compared with species from Italian Jurassic beds referred by Zigno[1] to the same genus, and with *S. robustus*, Newb.,[2] from the Rhætic of Honduras.

Text-fig. 12. (Specimen in the Oxford Museum.)

A single pinna, 9 cm. long and rather more than 4 cm. wide in the broadest part; the veins are numerous and spreading, differing both in their more divergent course and in their greater number from those of *Podozamites*.

V. 4069. Pl. XI. Fig. 4.

A single pinna 8 cm. long. The asymmetrical form of the lamina, the course and greater number of the veins, are characters by which

---

[1] Zigno (81), pls. xxxix.–xl.

[2] Newberry (88), p. 347, pl. viii. fig. 14.

we may distinguish this type from *Podozamites* as represented in Figs. 1 and 2, Pl. XI.

Stonesfield.

V. 3421. Leaflet 6 cm. × 1·7 cm.; blunt apex

Eyeford, Gloucestershire (Stonesfield Slate). *Brodie Coll.*

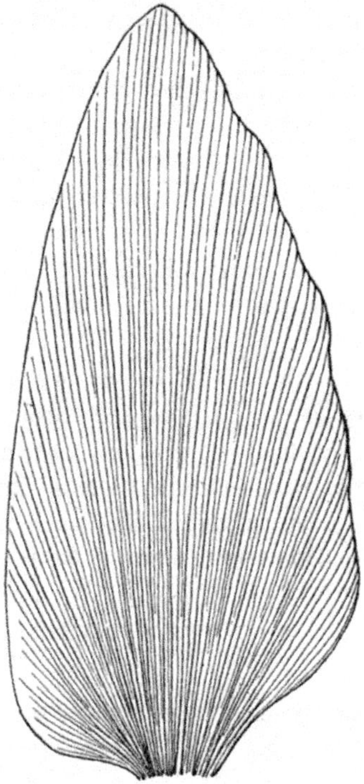

FIG. 12.—*Sphenozamites Belli*, sp. nov. Nat. size. Presented by Mr. Bell to the Oxford Museum.

V. 3436. Labelled "*Naiadea ovata.*" Similar to V. 4069 (Fig. 4), but the venation is less distinctly shown.

Eyeford. *Brodie Coll.*

Other specimens:—**V. 4642** (Stonesfield), **41,381** (Stonesfield, *Morris Coll.*), **52,817.**

A Genus of Gymnosperms of Doubtful Affinity.

Genus **PODOZAMITES**, Braun.

[Münster, Beit. Petrefact. Heft vi. p. 36, 1843.]

I have elsewhere suggested that *Podozamites* may be a genus of Conifers similar in habit to the recent *Agathis australis*, Salisb.[1] The Stonesfield specimens are detached leaves, or leaflets, and afford no additional evidence as to affinity; in shape and form they present a striking resemblance to the broader leaves of *Agathis*.

Podozamites stonesfieldensis, sp. nov.

(Pl. XI. Figs. 1–2; Pl. III. Fig. 4.)

1845. *Naiadea ovata*, Buckman, Geol. Cheltenham, p. 67, pl. ii. fig. 1.
Bensonia ovata, ibid. p. 67.
Lilia lanceolata, ibid. pl. ii. fig. 3.
1854. *L. lanceolata*, Morris, Brit. Foss. p. 11.
Naiadea ovata, ibid. p. 12.
? *Bensonia ovata*, ibid. p. 3.
1871. *Naiadea ovata*, Phillips, Geol. Oxford, p. 169.
Bensonia ovata, ibid. p. 168.
Lilia lanceolata, ibid. p. 168.
1886. *L. lanceolata*, Gardner, Geol. Mag. vol. iii. p. 203.
Bensonia ovata, ibid.
1894. *Lilia lanceolata*, Woodward, Lower Ool. p. 598.
Bensonia ovata, ibid.

I have ventured to designate the Stonesfield specimens represented in Pl. XI. Figs. 1, 2, and Pl. III. Fig. 4, by a new specific name, as there is some doubt as to the nature of the fossils figured by Buckman. It is practically certain that the specimen which he figures as *Naiadea ovata* is identical with this species, but the specific name *ovatus* has already been used in connection with the genus *Podozamites*. Possibly specimen No. V. 3468 (Pl. XI. Fig. 2) is the type of Buckman's *Naiadea ovata*, but the drawing is not sufficiently clear to render this certain.

---

[1] Seward (00), p. 241.

The specimen represented in Buckman's pl. ii. fig. 3 as *Lilia lanceolata* is, I have little doubt, specifically identical with *Podozamites stonesfieldensis*. The species named by Buckman *Bensonia ovata* (Geol. Chelt. p. 67) is not represented by a drawing; it is described as being characterised by "leaves oblongo-ovate, entire, with short stalk,"[1] and may be *Podozamites*. In a paper on Mesozoic Angiosperms Mr. Starkie Gardner alludes to the genera *Lilia* and *Bensonia* of Buckman, and speaks of them as being either "Cycadaceous or too indistinct to be determinable."[2]

Localities and Horizons.—Sevenhampton, Eyeford (Stonesfield Slate, Great Oolite).

The material is too fragmentary to admit of a satisfactory specific diagnosis. The leaves or leaflets (if the plant is a Conifer, allied to *Agathis*, the term leaf must be used; if it is Cycadean the term pinna or leaflet is more accurate) are symmetrical about a median axis, oblong-ovate in shape, entire, tapering somewhat abruptly to a narrow base and more gradually to the apex; veins rather more than 1 mm. apart, converging towards both the base and apex of the lamina.

Podozamites stonesfieldensis bears a fairly close resemblance to *Podozamites lanceolatus*, var. *latifolius*, as figured by Schenk,[3] from China, and to Geyler's Japanese species, *P. Reinii*.[4]

V. 3420. Pl. XI. Fig. 1.

Leaf 7·5 cm. long; the gap in the middle of the lamina has probably been produced by tearing. Veins very well defined, simple, or forked either close to the base or during their course through the lamina.

Eyeford. *Brodie Coll.*

V. 3468. Pl. XI. Fig. 2.

Possibly the original specimen figured by Buckman as *Naiadea ovata*; 8 cm. long by 3·6 cm. broad; similar to V. 3420 (Fig. 1), but more complete. The base is slightly notched, as if originally attached directly to an axis.

Sevenhampton. *Brodie Coll*

---

1 Buckman (45), p. 93.
2 Gardner (86), p. 203.
3 Schenk (83), pl. xlix. figs. 4*b*, 5.
4 Geyler (77), pl. xxxiii.; Yokoyama (89), pls. iii.–iv.

V. 4662. Pl. III. Fig. 4.

Possibly a smaller leaf or leaflet of this species; 2·1 cm. long by 1·2 cm. broad.

V. 3435. A specimen labelled "One of the Liliaceæ." Cf. Buckman's pl. ii. fig. 3.[1]

Sevenhampton. *Brodie Coll.*

Other specimens:—**V. 3437, V. 4070.**

GYMNOSPERMOUS SEEDS.

Numerous seeds have been obtained from various horizons and localities which it is hopeless to attempt to refer to specific types or to assign with confidence to their systematic position. On the other hand, there are many seeds which can be recognised by their shape as belonging to fairly well defined types; an example of a seed which may be designated by a specific name is *Carpolithes conicus* as figured by Lindley & Hutton. Until we find *C. conicus* with its tissues petrified or attached to a supporting organ we cannot determine its precise position; but, on the whole, it is more likely to belong to a Cycadean plant than to a Conifer. There is one difficulty that confronts us in dealing with such seeds as *Carpolithes conicus*, and that is the difference in appearance between a fairly complete seed and an internal cast; this is shown in the example from Malton represented in Text-fig. 15. In some instances it is possible to recognise with a fair degree of certainty detached seeds of Araucarian cones; in this case we are able to compare them with seeds or seed-impressions on cone-scales of the *Araucarites* type. In the following pages I have dealt with two sets of seeds under distinct specific names, *Carpolithes conicus* and *C. diospyriformis*, while several others are designated *Carpolithes*, sp. It is a hopeless and futile task to attempt to discriminate between isolated gymnospermous seeds which differ from one another chiefly in size. The generic name *Carpolithes* is used in a comprehensive sense for seeds of doubtful botanical position; in a former volume[2]

---

[1] Buckman (45).

[2] Seward (95), p. 101.

I have discussed at some length the application of this term to fossil seeds. It has been shown by Carruthers that some of the Great Oolite fossils referred to *Carpolithes* are more likely to be reptilian eggs.[1]

Genus CARPOLITHES, Sternberg.

[Flor. Vorwelt, iv. p. xl. 1823.]

Carpolithes conicus, Lindley & Hutton.

[Foss. Flor. pl. clxxxix. figs. 1, 2, 4, 1836.]

(Pl. XIII. Fig. 5; Text-figs. 13–15.)

1836. *Carpolithes conica*, Lindley & Hutton, Foss. Flor. pl. clxxxix. figs. 1, 2, 4.
C. Bucklandi, op. cit. pl. clxxxix. figs. 3, 5.
1845. *C. conicus*, Buckman, Geol. Cheltenham, p. 67, pl. ii. figs. 5–6.
C. Bucklandi, Unger, Synopsis, p. 188.
C. conica, ibid.
1848. *C. Bucklandi*, Bronn, Ind. Pal. p. 239.
C. conicus, ibid.
1849. *C. Bucklandi*, Brongniart, Tableau, p. 106.
C. conica, ibid.
Ulospermum ornatum, Pomel, Flor. foss. France, p. 16.
U. conicum, ibid.
1850. *Carpolithes Bucklandi*, Unger, Gen. spec. plant. foss. p. 309.
C. conica, ibid.
1854. *C. conicus*, Morris, Brit. Foss. p. 4.
1870. *Cycadinocarpus conicus*, Schimper, Trait. pal. vég. vol. ii. p. 210.
C. Bucklandi, ibid.
1878. *Carpolithus conicus*, Hudleston, Proc. Geol. Assoc. vol. v. p. 494.
1885. *Cycadeospermum Bucklandi*, Zigno, Flor. foss. Oolit. p. 160.
C. conicum, ibid.
1892. *Carpolithes Bucklandi*, Fox-Strangways, Jurass. Rocks, p. 302.
C. conicus, ibid.
1894. *Cycadinocarpus conicus*, Woodward, Lower Ool. p. 599.

The original specimen figured by Lindley & Hutton in their pl. clxxxix. fig. 4 as *Carpolithes conica* is in the Manchester Museum, and is redrawn in Text-fig. 13. Another seed in the same collection (361, Text-fig. 13*a*) may be the original of *Carpolithes Bucklandi* as represented in the *Fossil Flora*, pl. clxxxix. fig. 5.

---

[1] Carruthers (71[2]).

I have no hesitation in uniting *Carpolithes conicus*, L. & H., and *C. Bucklandi*, L. & H., *ex* Will. MS., under one specific name. The seeds designated by these names were obtained from the Coralline Oolite of Malton in Yorkshire.

The seeds grouped under *Carpolithes conicus* are characterised by their conical form, by the broad truncate surface, which is probably the base of the seed, traversed by either one or three prominent ridges. The margin of the truncate surface is usually provided with irregular tooth-like prominences, which were most probably pierced by vascular bundles which passed into the seed from its supporting organ. On some of the seeds the lateral surface bears numerous and irregularly disposed tubercles, a character which Lindley & Hutton mention as peculiar to *Carpolithes Bucklandi*. These lateral tubercles exhibit no uniformity either in their arrangement or in their occurrence, and they cannot, I think, be regarded as having a taxonomic importance. They may have been caused by the attacks of insects, and are probably not an original feature of the seeds. In some seeds, e.g. V. 9004 (Pl. XIII. Fig. 5), a single median ridge (Fig. 5, *a*) extends across the truncate end, while in others (Text-fig. 13*c* and Text-fig. 14 C) three ridges extend from the slightly raised centre of the truncate surface to the edge, giving the seed a somewhat triangular appearance. The difference in the form of the seeds, whether biconvex in section or triangular, cannot reasonably be considered a specific character. In the recent *Ginkgo biloba* (the maidenhair-tree) the seed, or rather the embryo-sac, is usually two-angled, but three-angled seeds are sometimes met with.[1] The peripheral teeth are clearly seen in A, B, and C, Text-fig. 14. In some examples the truncate surface is almost flat, with a central papilla (Text-fig. 13*b*), while in others (Text-fig. 13*a*) it is broadly conical. A specimen from the Malton Museum represented in Text-fig. 15 shows a seed cut through longitudinally; the interior is seen to be occupied by a cylindrical cast, bluntly terminated at one end and more pointed at the other, resembling the nut of a *Ginkgo* seed.

As regards the morphological nature of the parts of the seeds seen in the fossil specimens, the fairly thick outer envelope which forms

---

[1] Penzig (94), p. 515.

the surface in the seeds represented in Fig. 5, Pl. XIII., and in Text-figs. 13 and 14, was probably a woody integument, which, as in recent Cycad seeds and in the seeds of *Ginkgo*, was perhaps enclosed in a thick fleshy envelope formed from the more external portions of the integument. The internal cast seen in Text-fig. 15 may be regarded as replacing the embryo-sac. A comparison may be made between the seed shown in section in Text-fig. 15 and the drawing of a *Ginkgo* seed reproduced in a paper on the Maidenhair-tree published in 1900.[1]

In the seeds of some recent species of Cycads, e.g. *Macrozamia Fraseri*, there is a large smooth area at the base of the seed by which it was attached to the carpophyll, which is separated from the rest of the seed by a slightly projecting ridge formed by the fleshy portion of the integument.[2] This smooth basal area may be compared with the truncate surface of the fossil seeds.

It is possible that some of the specimens referred to as *Carpolithes*, sp., may be the internal casts of *Carpolithes conicus*. In all probability *Carpolithes conicus* was borne by a Cycadean plant, or perhaps by a member of the Ginkgoales; the point cannot be definitely settled without further evidence, but I incline to the view that the seeds are those of a Cycadean genus.

Numerous specimens of *Carpolithes conicus* may be seen in many museums, e.g., Jermyn Street, York, Scarborough, Whitby, Malton, and several others.

V. 9004. Pl. XIII. Fig. 5.

A conical seed, 3 cm. long, with truncate and slightly arched base (upper part in the drawing), traversed by a single median ridge, and surrounded at the margin by numerous teeth occurring either singly or in pairs. The breadth of the broad end is 2·2 cm.; the ridge shown in side-view in the figure, and passing over the middle of the truncate end of the seed, becomes somewhat broader in the centre, at *a*, probably due to the insertion of a vascular bundle.

Malton (Coralline Oolite).

---

[1] Seward & Gowan (00), pl. ix. fig. 45.

[2] Some seeds of this species in the Botanical Department of the British Museum, having a length of 5 cm. and a breadth of 3·5 cm., show these features very clearly. See Miquel (47), pl. iii.

Text-fig. 13. The seed shown in Fig. *a* may be the original of *Carpolithes Bucklandi*, fig. 5, pl. clxxxix., in the *Fossil Flora* of Lindley & Hutton. It differs from the more typical examples of *Carpolithes conicus* in its slightly larger size, in the more elevated

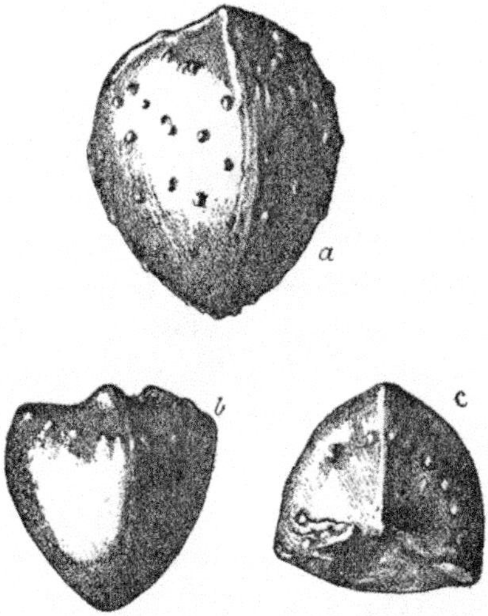

FIG. 13.—*Carpolithes conicus*, L. & H. From the type-specimens of Lindley & Hutton in the Manchester Museum. *a*, No. 361; *b*, *c*, No. 360. Nat. size.

base, and in the presence of numerous tubercles. The seed shown in side-view (*b*) and in basal view (*c*) is the original of *C. conicus* as figured by Lindley & Hutton (pl. clxxxix. fig. 4).

FIG. 14.—*Carpolithes conicus*, L. & H. Drawn from a specimen in the Malton Museum. A and C nat. size, B × 5.

Text-fig. 14. A well-preserved seed, 3 cm. long, drawn from a specimen in the Malton Museum; this example shows very clearly the three basal ridges (Fig. C) and the well-defined tooth-like projections (A and B).

Text-fig. 15. A seed from the Malton Museum showing the testa (integument) and the cast of the embryo-sac.

V. 9007. Pl. III. Fig. 3. An internal cast, probably of *C. conicus*, showing a single median ridge across the broad end.

Malton.

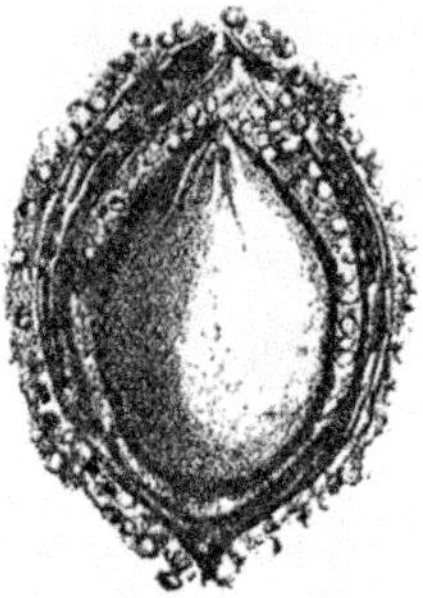

FIG. 15.—*Carpolithes conicus*, L. & H. Drawing of a specimen in the Malton Museum. Nat. size.

V. 9005. Several casts of seeds, some with a single median ridge, as in V. 9004 (Fig. 5, Pl. XIII.), others with three ridges meeting in the centre of the broad truncate end.

Other specimens :—**V. 9006**; **V. 9008**; **40,553** (Malton, *Bowerbank Coll.*); **40,580** (Kirby Moorside, *Bean Coll.*); **41,399** (*Morris Coll.*); **44,399** (*Mantell Coll.*); **52,604** (Malton, *Morris Coll.*); **52,938** (Malton, *presented by the Hon. Robert Marsham*, 1878).

Carpolithes diospyriformis, Sternberg.

[Flor. Vorwelt, iii. p. 37, pl. xxxvii. fig. 6, 1823.]

(Pl. XII. Fig. 7; Pl. XIII. Figs. 6–7.)

1823. *Carpolites diospyriformis*, Sternberg, Flor. Vorwelt, iii. p. 37, pl. xxxvii. fig. 6.
1825. *C. diospyriformis*, op. cit. iv. p. xl.
1845. *C. diospyriformis*, Unger, Synopsis, p. 256.
1848. *Carpolithes diospyriformis*, Bronn, Ind. Pal. p. 239.
1850. *C. diospyriformis*, Unger, Gen. spec. foss. plant. p. 517.
1854. *C. diospyriformis*, Morris, Brit. Foss. p. 5.
1871. *Carpolithus diospyriformis*, Phillips, Geol. Oxford, p. 171, Diag. xxxii. fig. 2.
1894. *Cycadinocarpus Lindleyanus* (pars), Woodward, Lower Ool. p. 599.

The seeds referred to *Carpolithes diospyriformis* differ sufficiently from those of the preceding species to be regarded as a distinct type. The example represented in Fig. 7, Pl. XIII. may be described as a typical seed of this species; it consists of a slightly asymmetrical pyriform cast, prolonged at one side of the broad base into a short stalk (*s.*), and tapering distally into an ill-defined apical region. This cast is frequently enclosed by an integument, represented either by a cavity in the rock-matrix or by a distinct shell, which appears to be absent from the base of the seed (Figs. 6–7, Pl. XIII.). The smaller seed shown in Fig. 6 may perhaps represent a distinct species, but so far as the material enables one to form an opinion, I am inclined to refer both to *C. diospyriformis*; the smaller cast has the same lateral area of attachment (*s.*) as occurs in the larger examples. The small cast represented in Fig. 7, Pl. XII. also shows very clearly the laterally situated basal scar (*s.*).

I am unable to express an opinion worth consideration as to the nature of the plant which bore *Carpolithes diospyriformis*. Specimens of *C. diospyriformis* are well represented in the Jermyn Street Museum, the Manchester Museum, the Sedgwick Museum, Cambridge, and in various museums in Yorkshire.

V. 2513. Pl. XIII. Fig. 7.

A cast 2·2 cm. in length, surrounded, except at the base, by a shallow groove—probably representing an integument—which appears to become continuous with the narrow apical neck of the cast. Slightly to the left of the middle of the broader (basal) end the cast is prolonged into a short stalk (*s.*).

Stonesfield. *Baber Coll.*

V. 2513*a*. Pl. XIII. Fig. 6.

A smaller and broader seed, possibly a distinct species, 1·2 cm. by 7 mm., of pea-shaped form, showing a lateral area of attachment or short stalk at *s*. *Baber Coll.*

V. 2513*b*. Pl. XII. Fig. 7.

A small internal cast with a lateral scar, *s*., probably specifically identical with V. 2513 (Pl. XIII. Fig. 7). *Baber Coll.*

V. 4648. A well-preserved specimen, similar to that shown in Fig. 7, Pl. XIII.

Other specimens:—**V. 82** (several seeds, *Egerton Coll.*); **V. 2513** (several seeds in addition to those figured, *Baber Coll.*); **V. 4649** (*R. Brown Coll.*); **V. 4663**; **V. 4671** (Sevenhampton); **V. 6347** (Stonesfield); **V. 6582**; **V. 6590**; **40,514** (*Bowerbank Coll.*); **40,581** (probably *C. diospyriformis*, from the Kellaways Rock, Scarborough, *Bean Coll.*); **41,168** (Stonesfield); **52,866** (several specimens, *S. Sharp Coll.*); **52,938***a* (Stonesfield, *presented by the Hon. Robert Marsham*, 1878).

Carpolithes, sp.

Under this designation are included several fossil seeds from Stonesfield and elsewhere that do not sufficiently resemble either of the preceding types to be definitely referred to *C. conicus* or to *C. diospyriformis*. The practice of appending specific names to casts of Gymnospermous seeds which do not exhibit any well-marked distinctive features seems to me to serve no useful purpose.

V. 224 (Malton, Coralline Oolite, *Morris Coll.*); **40,514** (*Bowerbank Coll.*); **40,581** (Kellaways Rock, Scarborough, *Bean Coll.*); **52,549** (Malton, *Wetherell Coll.*); **52,895** (*Toulmin Smith Coll.*); **52,938** (*presented by the Hon. Robert Marsham*, 1878).

Class CONIFERALES.

Family ARAUCARIEÆ.

Genus ARAUCARITES, Presl.

[Sternberg's Flor. Vorwelt, vii. p. 203, 1838.]

The species described below as members of the genus *Araucarites* are founded on fossil cones possessing features characteristic of the female shoots of recent species of *Araucaria*. The data at our disposal are not sufficiently complete to enable us to define with clearness or certainty the distinguishing features of the various examples of cones; the important fact is that fossil cones from several localities and from different sub-stages of the Jurassic system afford trustworthy evidence of the comparative abundance of Araucarian plants. It has long been recognised that the existing members of the Araucarieæ represent a dwindled remnant of a section of Coniferæ that was widely distributed in the northern hemisphere during the Jurassic and Wealden epochs. The evidence on which this statement is based is afforded by petrified wood exhibiting the well-marked histological characteristics of *Araucaria* and *Agathis*, and by cones composed of large single-seeded scales. It is intended to consider in detail the geological and geographical distribution of the Araucarieæ in a memoir on recent and fossil forms that is now in progress in collaboration with Miss Ford, of Newnham College, Cambridge.

The specific nomenclature adopted in the following pages may probably not represent differences worthy of specific rank, but we cannot hope to arrive at a satisfactory decision as to the limits of types without additional data.

Araucarites sphærocarpus, Carr.

[Geol. Mag. vol. iii. p. 249, pl. xi. 1866.]

(Pl. XIII. Figs. 2–4, 8.)

1866. *Araucaria sphærocarpa*, Carruthers, Geol. Mag. vol. iii. p. 249, pl. xi.
1870. *A. sphærocarpa*, Schimper, Trait. pal. vég. vol. ii. p. 254.
1871. *Araucarites sphærocarpus*, Carruthers, Geol. Mag. vol. viii. p. 543.
1872. *A. sphærocarpus*, Thiselton-Dyer, Geol. Mag. vol. ix. p. 151, fig. 4.
Cf. *A. Haeberleinii*, ibid. p. 150, figs. 1–3.
A. sphærocarpus, Balfour, Palæont. Bot. p. 62, figs. 83–85.
1878. *Araucaria sphærocarpa*, Lyell, Elements Geol. p. 335, fig. 351.
1884. *Araucarites sphærocarpa*, Saporta, Pal. Franç. vol. iii. p. 416, pl. lix. fig. 4.
1888. *Araucaria sphærocarpa*, Schenk, Handbuch, p. 171.
1894. *Araucarites sphærocarpus*, Woodward, Lower Ool. p. 599.

Locality and Horizon.—Bruton, Somersetshire (Inferior Oolite).

Araucarites sphærocarpus was founded by Carruthers on a large cone obtained from the Inferior Oolite of Somersetshire; he defined the species as follows:—"Cone spherical. Scales rhomboidal, with a central ridge produced into a stout, somewhat reflexed spine, and an obvious furrow dividing the scale into an upper and lower portion. Twenty to twenty-four scales in each spiral series in the centre of the cone." Each scale bears a single median seed as in the recent Araucarias; the scales have laterally expanded winged margins, and present a close agreement, both as regards the position and shape of the seed and in their form, with the seed-scales of the cones of *Araucaria Cookii*, *A. excelsa*, and other species belonging to the Eutacta division of *Araucaria*.

An important diagnostic character which should be added to Carruthers' definition is the occurrence of a single oval seed on each cone-scale, the seed being situated in the median line of the proximal portion of the scales.

Type-specimen of Carruthers in the British Museum (41,036; Pl. XIII. Figs. 2–4, 8). The Jermyn Street Museum and the British Museum possess plaster-casts of Carruthers' 'type.'

41,036. Pl. XIII. Figs. 2–4, 8.

Type-specimen of Carruthers (Geol. Mag. pl. xi. 1866).

The specimen represents part of a spherical cone approximately 13 cm. in length and 13 cm. broad; a small piece of the peduncle is preserved. The cone-scales, especially in the upper part of the cone, have been somewhat worn down distally, but they are more perfectly preserved near the base. The exposed end of a cone-scale measures 2 cm. across; it is divided by a transverse groove (*d*, Figs. 3–4) into two portions, a smaller upper and a larger lower portion. Near the upper edge of the lower half a row of small pits is clearly shown (Fig. 3) which no doubt represent vascular bundles exposed on the worn-down surface of the cone-scales. Fig. 4 shows the scales in a more complete state; the pits are faintly indicated, and a central umbo is seen on the lower part of each scale.

In Fig. 2, Pl. XIII. a scale is shown cut through longitudinally, exposing the seed, *s.*, embedded in the substance of the scale tissues, as in recent Araucarias. Fig. 8 affords a surface-view of a detached scale; the slightly depressed area *s.* shows the position formerly occupied by the seed, and at *d* one sees the transverse groove dividing the distal portion of the scale into an upper and a lower half (compare Fig. 4, *d*). The umbo, seen in surface-view in Fig. 4, is shown in side-view at *u* in Fig. 8.

Bruton, Somersetshire (Inferior Oolite). *Purchased*, 1861.

V. 6373. An artificial cast of the type-specimen (41,036).

Araucarites ooliticus (Carruthers).

[Geol. Mag. vol. v. p. 156, pl. ix. figs. 1–6, 1868.]

(Pl. XIII. Fig. 1; Text-figs. 16–18.)

1868. *Kaidacarpum ooliticum*, Carruthers, Geol. Mag. vol. v. p. 156, pl. ix. figs. 1–6.
1872. *K. ooliticum*, Balfour, Palæont. Bot. p. 84, fig. 88.
1873. *Pandanocarpum ooliticum*, Zigno, Flor. foss. Oolit. vol. ii. p. 3.
Kaidacarpum ooliticum, Sharp, Quart. Journ. Geol. Soc. vol. xxix. p. 290.
1877. *Araucarites Hudlestoni*, Carruthers, Quart. Journ. Geol. Soc. vol. xxxiii. p. 402, pl. xvii.
1878. *Araucaria Hudlestoni*, Hudleston, Proc. Geol. Assoc. vol. v. p. 494.
1886. *Kaidacarpum ooliticum*, Gardner, Geol. Mag. vol. iii. p. 199.
1888. *Araucaria Hudlestoni*, Schenk, Handbuch, p. 171.
1890. *A. Hudlestoni*, Zittel, Handbuch, p. 280.
1892. *A. Hudlestoni*, Fox-Strangways, Jurass. Rocks, p. 334.
1894. *Pandanocarpum (Kaidacarpum) ooliticum*, Woodward, Lower Ool. p. 317.
1895. *Araucarites Hudlestoni*, Seward, Wealden Flor. ii. p. 191.
1896. *Kaidacarpum ooliticum*, Seward, Annals Botany, vol. x. p. 214.

Localities and Horizons.—Kingsthorpe, Northants (Great Oolite); Malton, Yorkshire (Coralline Oolite).

The type-specimen of *Araucarites ooliticus* (*Kaidacarpum ooliticum*, Carr.) is in the Northampton Museum; the type of Carruthers' species *Araucarites Hudlestoni* is in the York Museum.

The genus *Kaidacarpum* was instituted by Carruthers in 1868, and defined as follows:—"Fruit composed of pyramidal rhomboidal single-seeded drupes, sessile or sub-sessile on a thickened spadix."[1] I have elsewhere[2] expressed the opinion that the specimens included by Carruthers in *Kaidacarpum*, and regarded as Mesozoic examples of the Monocotyledonous family Pandaneæ, are cones of Araucarian plants. This view has been, in my opinion, confirmed by an examination of the type-specimen of *Kaidacarpum ooliticum* in the Northampton Museum. The specimen represents a portion of a cone 9 cm. long, consisting of a cylindrical projecting central portion covered with spirally disposed deep pits or meshes bounded by a framework of crystalline ridges; attached laterally to this central region, and partially covered

---

[1] Carruthers (68), p. 153.
[2] Seward (96), p. 213.

by the matrix of the rock, are numerous imbricate scales. On one side of the central region one sees that some of the pits or pockets are occupied by seeds which extend distally into the hollow basal portion of the cone-scales. The scales are approximately 1·7 cm. broad and slightly winged. At the base of the cone are portions of small sterile scales. The features presented by this cone agree precisely with those of recent Araucarian cones; the single seed embedded in each scale, the general form of the cone, the form of the individual cone-scales, and the occurrence of sterile scales at the base are characters which can be exactly matched in recent Araucarias. I have included the species *Araucarites Hudlestoni*, Carr., under *A. ooliticus*, as I fail to discover any differences that appear to be of specific rank.

Another specimen in the Northampton Museum, drawn in Text-fig. 17, shows a portion of a smaller cone: in A a scale is represented in surface-view; the thin lateral borders are clearly shown in both A and B, Text-fig. 17; from the lower margin of the scale (A) the end of a seed projects, and in the end-view (B) one sees the cavity originally occupied by a seed. The example shown in Text-fig. 18, also from Kingsthorpe, illustrates the appearance presented by a small cone as seen in surface-view.

Araucarites ooliticus bears a fairly close resemblance to a French cone of Lower Cretaceous age described by Fliche as *Sarcostrobilus Paulini*,[1] and referred by him to the Araucarieæ.

The specimen figured by Lindley & Hutton[2] as *Strobilites Bucklandi* is no doubt an Araucarian cone, possibly identical with *Araucarites ooliticus*, but no information is given as to the geological age.

The name *Araucarites Hudlestoni* was instituted by Carruthers for a cone from the Coralline Oolite of Malton; it is defined as follows:—

"Cone oblong-ovate, supported on a thick branch, which is clothed with leaves to the base of the cone. Scales numerous, supported on a thick axis. Scales small, wingless, with a well-marked lepidium or upper scale. Seed small, oval, borne at the base of the scale."

---

[1] Fliche (00), p. 11, pl. i. figs. 2–5.

[2] Lindley & Hutton (34), pl. cxxix.

In one of the specimens in the York Museum described by Carruthers the cone is seen in longitudinal section; a broad central region, representing the stout cylindrical axis, is occupied by a mass of pisolite to which numerous cone-scales are attached. The scales are shown in section, and in several of them one sees a single seed lying in a cavity occupying the proximal end of the

FIG. 16.—*Araucarites oolitieus* (Carr.). Surface-view of part of the mould in which the type-specimen (cast) of *Araucarites Hudlestoni*, Carr., was embedded. From a specimen in the York Museum. Nat. size.

scale, precisely as in *Araucarites ooliticus* from Northamptonshire. At the apex of the specimen the vascular cylinder of the cone-axis is seen in section, but the minute structure is not preserved. Another specimen represents the pedunculate base of the cone in which the scales are seen in end-view; they reach a breadth of 1·4 cm., and may fairly be described as winged. Text-fig. 16 shows

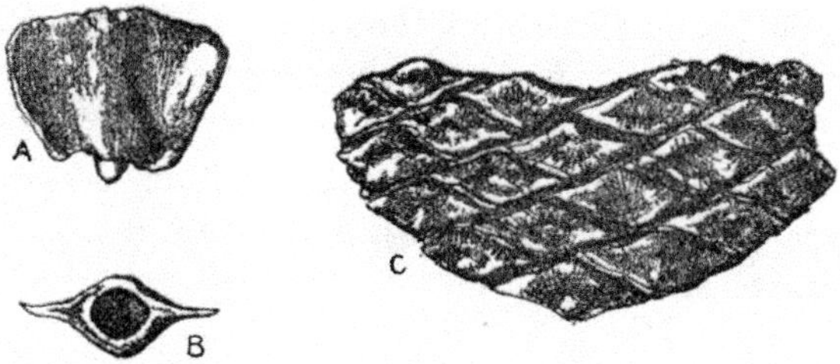

FIG. 17.—*Araucarites ooliticus* (Carr.). A, scale in surface-view, with end of seed projecting; B, view of proximal end of scale; C, surface-view of cone. From specimens in the Northampton Museum. Nat. size.

a few of the distal ends of cone-scales seen on the mould made by the cone in the oolitic rock: a slight ridge extends across the end of each cone-scale, as in *Araucarites sphærocarpus* (Pl. XIII. Figs. 3–4).

It is possible that *Araucarites sphærocarpus* may be specifically identical with *A. ooliticus*, but it is probably wiser to retain both specific names. I have, however, no hesitation in regarding them as very closely allied forms belonging to the section Eutacta of

Araucaria, as represented by *Araucaria excelsa, A. Cookii* among recent species.

52,840. Pl. XIII. Fig. 1.

Part of a cone seen from near the proximal ends of the scales. The cavities in the scales were originally occupied by seeds. The scales themselves are approximately 1·6 cm. broad at the broadest end, but their distal faces are not exposed in the specimen; they are winged and closely resemble those of *Araucarites sphærocarpus* and *A. Brodiei*. *S. Sharp Coll.*

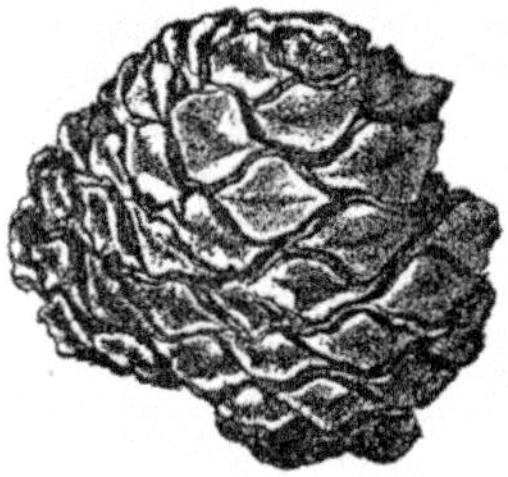

FIG. 18.—*Araucarites oolitiscus* (Carr.). From a specimen in the Northampton Museum. Nat. size.

V. 3465. An impression of the surface of a cone, showing the distal ends of the scales similar to those represented in Text-fig. 18.

Near Leckhampton. *Purchased,* 1894.

V. 4673. Section of a cone in a coarse oolitic rock, very similar to the type-specimen of *Araucarites Hudlestoni.*

Northamptonshire.

Araucarites Brodiei, Carruthers.

[Geol. Mag. vol. vi. p. 3, pl. ii. figs. 1–6, 1869.]

(Pl. XII. Fig. 2; Pl. III. Fig. 5.)

1869. *Araucarites Brodiei*, Carruthers, Geol. Mag. vol. vi. p. 3, pl. ii. figs. 1–6.
1870. *Araucaria Brodiei*, Schimper, Trait. pal. vég. vol. ii. p. 253.
1871. *Araucarites Brodiei,* Phillips, Geol. Yorks. p. 171, Diag. xxxii. fig. 4.
1872. *A. Brodiei*, Balfour, Palæont. Bot. p. 83, pl. ii. fig. 10.
1884. *Araucaria Brodiei*, Saporta, Pal. Franç. vol. iii. p. 416, pl. lix. figs. 5–6. Cf. *A. moreauana*, ibid. pl. clxxxiv.
1888. *A. Brodiei*, Schenk, Handbuch, p. 171.
1890. *A. Brodiei*, Zittel, Handbuch, p. 280.
1894. *Araucarites Brodiei,* Woodward, Lower Ool. p. 599.

Type-specimens in the British Museum (V. 3363, V. 3364)—one of them redrawn (V. 3364) in Fig. 2, Pl. XII.—and in the Oxford Museum. *Araucarites Brodiei* is thus defined by Carruthers:—

"Scales from the centre of the cone cuneate, gradually tapering towards the narrow base of attachment to the axis, composed of two portions, each terminating at its free apex in a short spinous process, the lower and larger portion very broad and membranous, the upper portion narrower and somewhat parallel-sided, supporting between them a single ovoid seed."

The material on which this species was founded is less complete than that which was made the type of the two preceding species: it is indeed doubtful whether the designation *Araucarites Brodiei* serves a useful purpose. The cone-scales agree closely in form with those already described, and possess a winged border.

The single scale represented in Fig. 5, Pl. III. is drawn from an unusually well-preserved specimen from Stonesfield in the Manchester Museum: the scale is 3 cm. long and 1·7 cm. broad, bearing a seed approximately 1·4 cm. in length. The raised ridge just beyond the distal end of the seed or seed-cavity no doubt corresponds to the so-called ligule characteristic of the cone-scales of some recent species of *Araucaria*. The spinous process from the broad end of the scale may be compared with a similar appendage borne by the large cone-scales of *Araucaria Bidwilli*, Hook.

We may perhaps include this scale under *Araucarites Brodiei* as probably having been derived from a larger cone of that species than were the scales figured by Carruthers in his original account of the type.

Pl. III. Fig. 5. A cone-scale in the Manchester Museum, showing distal spinous process, ligule, and a portion of a seed.

Stonesfield. *Manchester Museum Coll.*, No. 197.

V. 3364. Pl. XII. Fig. 2.

An incomplete cone-scale (figured also by Carruthers, Geol. Mag. pl. ii. fig. 3, 1869), 2 cm. long by 1·8 cm. broad, with part of a seed shown as a projecting cast lying in a depression. There is a distinct depression in the middle of the distal end of the scale, a character shown more clearly in the specimen represented in Pl. III. Fig. 5 (Manchester Museum). Several other scales occur on the same piece of rock, with fragments of *Thuites expansus*.

Brodie Coll.

V. 3363. Figured by Carruthers, Geol. Mag. pl. ii. fig. 1, 1869. This specimen is difficult to interpret; it appears to consist of the impression of a stalk bearing rhomboidal leaf-areas, which are not quite so distinctly marked as in Carruthers' figure. The stalk expands into what is probably the base of a large cone, but the preservation is too imperfect to admit of a more detailed description. There is no positive evidence that the cone of which a portion is represented by the specimen bore scales identical with the detached cone-scales referred by Carruthers to *Araucarites Brodiei.*

Brodie Coll.

Other specimens:—**V. 6593, V. 9707, 41,384, 41,404** (a large slab with several cone-scales of *Araucarites*, also twigs and well-preserved cones of *Thuites expansus*; *Morris Coll.*).

Araucarites sphæricus (Carruthers).

[Geol. Mag. vol. iv. p. 105, pl. vi. fig. 8, 1867.]

1867. *Cycadeostrobus sphæricus*, Carruthers, Geol. Mag. vol. iv. p. 105.
1872. *Zamiostrobus sphæricus*, Schimper, Trait. pal. vég. vol. ii. p. 202.
1885. *Z. sphæricus*, Zigno, Flor. foss. Oolit. vol. ii. p. 151.
1895. *Cycadeostrobus sphæricus*, Woodward, Lower Ool. p. 402.

The cone which Carruthers described as *Cycadeostrobus sphæricus* is in all probability another example of *Araucarites*, but without seeing a specimen in section one cannot arrive at any definite conclusion.

Carruthers' type-specimen was found in the Oxford Clay, Wiltshire.

Araucarites Bucklandi (Lindley & Hutton).

[Foss. Flor. vol. ii. pl. cxxix. 1834.]

1834. *Strobilites Bucklandi*, Lindley & Hutton, Foss. Flor. vol. ii. pl. cxxix.
1850. *S. microphylla*, Unger, Gen. spec. foss. plant. p. 301.
1854. *S. Bucklandi*, Morris, Brit. Foss. p. 23.
1896. *Araucarites Bucklandi*, Seward, Annals Bot. vol. x. p. 216.

The specimens referred by Lindley & Hutton to *Strobilites Bucklandi* are no doubt portions of an Araucarian cone; the scales are shown to be hollow, and agree closely with those of *Araucarites oolíticus* (Carr.). The drawings published in the *Fossil Flora* were prepared for Dr. Buckland, who communicated them to Lindley & Hutton, but we have no information as to the locality or the geological horizon from which the fossils were obtained. In all probability the specimens are of Jurassic age, and form part of a cone closely allied to or identical with *Araucarites ooliticus*.

Araucarites Cleminshawi, Mansel-Pleydell.

[Proc. Dorset Nat. Hist. Antiq. Field Club, vol. v. p. 141, 1885.]

Under this name Mr. Mansel-Pleydell described an imperfect specimen of a spherical cone, with rhomboidal scales, from the Inferior Oolite of Sherborne, Dorsetshire. The cone is too imperfect to describe in detail, or to refer with any degree of certainty to a specific position. Possibly it may be identical with *Araucarites ooliticus*, but the data are too meagre to afford satisfactory evidence as to its relation to other forms; it is doubtless an Araucarian cone very similar to the other Jurassic examples already described.

Araucarites, sp.

SEEDS.

(Pl. XII. Fig. 6.)

1836. *Carpolithes*, Lindley & Hutton, Foss. Flor. vol. iii. pl. cxciii. fig. A 3.
1845. *C. conicus* (pars), Buckman, Geol. Cheltenham, pl. ii. fig. 5c.
1871. *Carpolithus Lindleyanus*, Phillips, Geol. Oxford, p. 173, Diag. xxxii. fig. 1.

Amongst the various seeds referred by authors to *Carpolithes*, there are some which may be identified with a fair degree of probability as Araucarian. Fig. A 3 of pl. cxciii. of the *Fossil Flora* of Lindley & Hutton, drawn from a Stonesfield specimen, probably represents an Araucarian seed detached from its scale. The term *Carpolithes Lindleyanus* has been applied by Phillips and others to seeds of which some are no doubt those of Araucarian cones; the same name has been used also by Dunker and Schimper for Wealden seeds[1] of a distinct type. I do not propose to apply a specific name

---

[1] Schimper (70), ii. p. 210.

to the seeds from Stonesfield and elsewhere which are regarded as Araucarian; it is impossible to refer them with certainty to a particular species.

V. 6591. Pl. XII. Fig. 6.

A detached seed, 1·2 cm. long, probably from an Araucarian cone. Stonesfield? *Robt. Brown Coll.*

Other specimens:—**V. 3443** (*Brodie Coll.*); ? **V. 3444** (*Brodie Coll.*); **V. 6347.** (Stonesfield.)

Family CUPRESSINEÆ?

The species *Thuites expansus* usually included in this family agrees with the recent members in the whorled arrangement of the leaves, but the structure of the cones bears a closer resemblance to that which is met with in the genus *Sequoia* and other conifers than those belonging to the Cupressineæ. Our knowledge of the fertile shoots of *Thuites* is too meagre to enable us to compare them in detail with the cones of existing types.

Genus **THUITES**, Brongniart.

[Tableau, p. 71, 1849.]

The customary inclusion of the comprehensive fossil genus *Thuites* in the Cupressineæ is open to question on the grounds that our knowledge is very incomplete as regards certain features of taxonomic importance. The family as represented at the present day is characterised by the arrangement of both vegetative and reproductive leaves in alternate whorls: the scale-leaves of *Thuites*, e.g. *T. expansus*, usually appear to conform in this respect to the similar leaves of *Cupressus*, *Thuja*, and other recent genera, but in some examples the leaves are by no means obviously whorled. This departure from the whorled disposition, which may be apparent and not an original character, need not prove an obstacle to the inclusion of *Thuites* in the Cupressineæ, as one occasionally notices the same tendency towards a spiral leaf-arrangement on recent twigs. A much more important discrepancy between the fossil and recent forms is noticed in the reproductive shoots. The female flowers of *Cupressus*, *Thuja* (including *Biota*), *Libocedrus*, *Callitris*, *Actinostrobus*, *Thujopsis* are characterised by the whorled arrangement and the peculiar valve-like or peltate form of the woody scales; the flowers of *Juniperus*, on the other hand, are distinguished by the fleshy or berry-like character of the paired cone-scales. The cones of *Thuites expansus*, on the other hand, as shown in Fig. 4, Pl. IX. and in the drawings given by Phillips,[1] consist of more numerous scales with a spiral disposition, and bear a much closer resemblance to those of the Taxodiinæ: we have unfortunately no satisfactory data as regards the number of seeds borne on each scale in the fossil cones. Professor Zeiller[2] has also drawn attention to the similarity between cones of *Thuites* and those of certain recent species of Taxodiinæ, and he expresses the opinion that the reference of this fossil genus to the Cupressineæ is by no means satisfactorily established.

It is probable that Don's genus *Athrotaxis*, represented by three existing species confined to Tasmania, may be regarded as a survival from the Mesozoic period; the vegetative branches of *A. cupressoides*, Don, bear a striking resemblance to the fossil genus *Brachyphyllum*, as well as to some of the forms referred to *Thuites*. The female flowers of the Tasmanian conifer appear to be more closely allied to those of *Thuites* than are the cones of recent Cupressineæ. Several Jurassic conifers have been referred to *Athrotaxites* of Unger[3] and to Fontaine's genus *Athrotaxopsis*,[4] and it is possible that in *Thuites expansus* also we have a type closely allied to *Athrotaxis*. In the present state of our knowledge it is impossible to arrive at a definite conclusion as to the affinity of *Thuites*; the important point is to recognise the insufficiency of the evidence which has led to the recognition of the fossil species as members of the Cupressineæ. It is at least clear that in *Thuites expansus* we have a very common and widely-spread Jurassic conifer in the northern hemisphere, which demonstrates the former abundance of a type in a region which no longer affords any species which can be regarded as closely allied to *Thuites expansus*.

---

[1] Phillips (71), p. 173, Diag. xxxii. fig. 5.
[2] Zeiller (00), p. 273.
[3] Dyer (72).
[4] Fontaine (89), p. 239, pls. cxiii.–cxvii.

Thuites expansus, Sternberg.

[Flor. Vorwelt, ii. p. 38, pl. xxxviii. figs. 1–2, 1823.]

(Pl. IX. Figs. 1, 4; Text-fig. 19.)

1823. *Thuites expansus*, Sternberg, Flor. Vorwelt, . p. 38, pl. xxxviii. figs. 1–2.
T. articulatus, ibid. p. 39, pl. xxxiii. fig. 3.
T. cupressiformis, ibid. p. 39, pl. xxxiii. fig. 2.
T. divaricatus, ibid. p. 39, pl. xxxix.
1828. *T. expansus*, Brongniart, Prodrome, p. 109.
T. (?) *divaricatus*, ibid.
T. (?) *cupressiformis*, ibid.
T. (?) *acutifolia*, ibid.
1838. *Caulerpites expansus*, Sternberg, Flor. Vorwelt, vii. p. 22.
C. thuiæformis, ibid.
C. Bucklandianus, ibid.
1845. *Thuites expansus*, Unger, Synopsis, p. 190.
T. divaricatus, ibid.
T. (?) *acutifolia*, ibid.
T. expansus, Buckman, Geol. Cheltenham, p. 67, pl. i. fig. 6.
T. cupressiformis, ibid. p. 67.
1848. *T. expansus*, Bronn, Ind. Pal. p. 1271.
T. divaricatus, ibid.
T. articulatus, ibid.
1849. *T.* (?) *expansus*, Brongniart, Tableau, p. 106.
T. divaricatus, ibid.
Brachyphyllum acutifolium, ibid.
Moreauia expansa, Pomel, Flor. jurass. p. 21.
M. divaricata, ibid.
M. acutifolia, ibid.
1850. *Caulerpites expansus*, Unger, Gen. spec. plant. foss. p. 6.
C. Bucklandianus, ibid.
Thuites expansus, Göppert, Foss. Conif. p. 182.
T. divaricatus, ibid.
T. articulatus, ibid. p. 183.
1854. *T. expansus*, Morris, Brit. Foss. p. 24.
T. divaricatus, ibid.
T. cupressiformis, ibid. p. 23.
T. articulatus, ibid.
1866. *T. expansus*, Carruthers, Geol. Mag. vol. iii. p. 545.
T. divaricatus, ibid.
T. cupressiformis, ibid.
T. acutifolius, ibid.

1870. *Echinostrobus expansus*, Schimper, Trait. pal. vég. vol. ii. p. 333.
1871. *Thuites expansus*, Phillips, Geol. Oxford, p. 171, Diag. xxxi. figs. 4–5.
T. divaricatus, ibid. figs. 7–8.
T. cupressifolius, ibid. p. 171.
T. articulatus, ibid. fig. 1.
1876. ? *Echinostrobus expansus*, Feistmantel, Gondwana Flor. p. 60, pl. ix.
1877. *E. expansus*, ibid. p. 17, pl. xi.
1884. *Palæocyparis expansa*, Saporta, Pal. Franç. vol. iii. p. 600, pl. lxxxi.
Cf. *P. Itieri*, ibid. pl. lxxvii.
1894. ? *Thuyites pulchelliformis*, Saporta, Flor. Portugal, pl. ix. fig. 5.
? *Palæocyparis flexuosa*, ibid. pls. xix.–xx.
Echinostrobus expansus, Woodward, Lower Ool. p. 599.
Thuites divaricatus, ibid.
T. cupressiformis, ibid.

The habit of the *vegetative shoots* agrees with that of *Brachyphyllum mamillare*, Brongn., and with recent species of *Cupressus* and *Thuja*; in some branches the smaller branchlets are crowded together (Pl. IX. Fig. 1), and in others (Text-fig. 19) the habit is more open. Leaves whorled, two in each whorl; appressed and decurrent on the axis, broad, slightly falcate, the apex free and broadly triangular in surface-view.

Female flowers in the form of more or less globular cones (Pl. IX. Figs. 4, 4*a*) consisting of a central axis bearing numerous spirally arranged cone-scales. No seeds were found in any of the specimens.

Male flowers longer and narrow, composed of numerous crowded sporophylls attached at right angles to the axis; each sporophyll is expanded distally, and bent upwards and downwards into a peltate lamina.

Localities and Horizons.—Practically all the specimens are from the Stonesfield Slate of Stonesfield, Eyeford, and Sevenhampton. The Jermyn Street Museum Collection includes a cone from the Forest Marble of Wolverton, which appears to be indistinguishable from that represented in Pl. IX. Fig. 4.

The most abundant fossil plants from the Stonesfield Slate are the numerous fragments of coniferous twigs referred by Sternberg and many subsequent writers to different species of the genus *Thuites*. Water-worn cones (female flowers) and male flowers, which undoubtedly belonged to the same plant as the vegetative shoots, are also fairly common. An examination of a considerable number of specimens of *Thuites* branches has convinced me that it is impossible to retain the specific names *Thuites expansus*, *T. divaricatus*, *T. articulatus*, and *T. cupressiformis* as designations of distinct species; it is by no means unlikely that among the numerous specimens we have more than one species represented, but I am unable, with the available data, to recognise any characters on which it is possible to found distinctive diagnoses. I have therefore adopted the plan of using Sternberg's name *Thuites expansus* as a comprehensive designation for the coniferous twigs from Stonesfield which agree more or less closely with the example represented in Pl. IX. Fig. 1. It has been pointed out by some previous writers that too many specific titles have been employed. Saporta,[1] for example, includes under *Palæocyparis expansa* the species *T. expansus*, *T. articulatus*, and *T. divaricatus*. Some of the drawings published by Saporta were made from specimens in the Oxford Museum, and Sternberg's figures are also taken from the same source. While most authors adopt the generic name *Thuites*, which implies a decussate arrangement of the scale-leaves, as in recent Cupressineæ, a few have substituted other designations. Schimper[2] notices that the leaves of *Thuites expansus* are not always in opposite pairs, and substitutes the generic name *Echinostrobus*.

Specimens of *Thuites expansus* are abundantly represented in several museums; in the Manchester Collection there is a very well preserved branch (No. 265) in which the whorled arrangement of the broadly triangular leaves is clearly shown. One of the best specimens of a female cone is in the Sedgwick Museum, Cambridge, but none of the examples are sufficiently well preserved to enable one to determine the structure of the seminiferous scales. Another fairly good example is the original of Phillips' figure (published in the *Geology of Oxford*, Diag. xxxii.), preserved in the Oxford Museum; this shows the form of the cone-scales, as seen in side-view, more clearly than most specimens. An imperfect cone from the Forest Marble of Wolverton in the Jermyn Street Museum is probably specifically identical with the Stonesfield type. The best example of a male flower that I have seen is in the collection of

---

[1] Saporta (84), p. 600.
[2] Schimper (70), p. 333.

Mr. A. M. Bell of Oxford; it is 1·4 cm. in length, and tapers gradually towards the apex. Each sporophyll, attached almost at right angles to the axis, terminates distally in a peltate expansion. A few oolitic grains occur in close association with the sporophylls, and at first sight these might be mistaken for pollen-sacs.

The specimen of a male flower figured by Phillips (Geol. Oxford, Diag. xxxii.) is in the Oxford Museum.

40,513. Pl. IX. Fig. 1.

A well-preserved impression in sandstone showing the habit of a vegetative shoot of *Thuites expansus* which agrees closely with that

FIG. 19.—*Thuites expansus*, Sternberg. V. 81*a*. Nat. size.

of recent Cupressineæ. The arrangement of the appressed leaves on the main axis is by no means obviously spiral, but on the lateral branches the leaves are clearly attached in pairs: those on the main axis are longer than those on the smaller axes; the smaller leaves are triangular in form with free apices.

Stonesfield Slate. *Bowerbank Coll.*

V. 81*a*. Text-fig. 19.

A less distinct impression in oxide of iron, but interesting as an example of the habit of the vegetative shoots in which the lateral branches are long and slender. Compare the recent *Biota orientalis*.
Stonesfield. *Egerton Coll.*

V. 2514. A specimen showing the branching habit clearly.
Stonesfield. *Baber Coll.*

V. 3432. Similar to the specimen figured by Buckman,[1] but probably not the actual one represented in the drawing.
Eyeford (Stonesfield Slate). *Brodie Coll.*

38,931. Larger specimen, 12 cm. long, in habit very similar to *Brachyphyllum mamillare*, Brongn.[2]
Stonesfield.

40,512. A branch 13 cm. in length; the leaves, in whorls of two, are clearly shown, and the deep moulds in the rock afford evidence of their thick lamina. *Bowerbank Coll.*

40,698. A fragment labelled *Cryptomerites divaricatus*; probably *Thuites expansus*.
Stonesfield. *Purchased*, 1859.

52,937. A good example of crowded branchlets, very like species of *Cupressus*. Note the contrast between specimens of this type and that shown in Text-fig. 19 (V. 81*a*).
Presented by the Hon. Robert Marsham, 1878.

Other specimens:—**V. 81** (several specimens); **V. 2515**; **V. 2631**; **V. 2935**; **V. 2936**; **V. 3365**; **V. 3417**; **V. 3418**; **V. 3432**; **V. 3458**; **V. 3637**; **V. 4072**; **V. 4073**; **V. 4677–4683**; **V. 4685**; **V. 4686**; **V. 4689–4691**; **V. 4693–4697**; **V. 4699**; **V. 6580**; **V. 6581**; **V. 6584**; **V. 6588**; **V. 6589**; **V. 7612**; **V. 7917**; **V. 9710**; **V. 9711**; **41,169**; **41,170**; **41,372**; **41,373**; **41,382**; **41,384**; **52,817**. (*Egerton, Beckles, Baber, Brodie, Buckman, Morris, Bright,* and *Slatter Colls.*)

---

[1] Buckman (45), pl. i. fig. 6*c*.
[2] Seward (00), p. 297, pl. x. fig. 1.

CONES (Flowers).

1. *Female Flowers.* **V. 3439.** Pl. IX. Figs. 4, 4*a*.

A small globular cone, considerably worn. The cone-scales appear to have a spiral arrangement; each scale (Fig. 4*a*, × 5) is occupied near one edge by a funnel-shaped cavity, the upper side of the funnel being characterised by several converging ridges. The water-worn nature of the specimen renders impossible an accurate description of the cone. A similar specimen is figured by Buckman.[1]

Sevenhampton. *Brodie Coll.*

41,382*a*. This is one of the few specimens showing a cone, or a portion of a cone, attached to a shoot. The cone (female) is clearly identical specifically with that shown in Fig. 4, Pl. IX.; part of the cone-axis is exposed, showing spirally disposed holes which were no doubt originally occupied by vascular bundles. The vegetative portion of the specimen is imperfectly preserved, and might be referred either to *Brachyphyllum* or *Thuites*. *Morris Coll.*

Other specimens:—**V. 4696** (an imperfect cone seen in section); **V. 6578**; **V. 6587.**

2. *Male Flowers.*

52,937. A small and imperfect male cone, showing a slender axis giving off numerous sporophylls at right angles; each sporophyll is somewhat enlarged at the distal end.

Presented by the Hon. Robt. Marsham, 1878.

V. 3441. A cone 2 cm. long and 1 cm. broad, shown in section. This specimen appears to be a male cone larger than most examples, but the preservation is very imperfect.

Stonesfield. *Brodie Coll.*

Other specimens:—**V. 3445** (*Brodie Coll.* Some of the sporophylls fairly clearly shown in section); **V. 6592**; **52,937***a*.

Thuites, sp.

(Text-fig. 20.)

52,838. The specimen shown in the figure was obtained from the Oxford Clay of Christian Malford, Wiltshire; it consists of

---

[1] Buckman (45), pl. i. figs. 6*a*-*b*.

a carbonaceous impression of a twig bearing thick fleshy leaves apparently disposed in pairs, but in some places (as in examples of

FIG. 20.—*Thuites*, sp. 52,838. Nat. size.

Thuites expansus) the whorled arrangement is by no means obvious.
Christian Malford (Oxford Clay). *Cunnington Coll.*

CONIFERALES INCERTÆ SEDIS.

Genus BRACHYPHYLLUM.

Brachyphyllum, sp. *a*.

(Cf. *B. mamillare*, Brongniart.)

(Pl. IX. Fig. 5.)

11,130. Pl. IX. Fig. 5.

This specimen may be provisionally referred to the genus *Brachyphyllum*; it agrees closely in habit with *Thuites expansus*, but the leaves are apparently spiral in their arrangement. This and other similar examples are too small and imperfect to assign with confidence to a distinct species, but they may be identical with the Yorkshire Inferior Oolite type, *Brachyphyllum mamillare*, Brongn. It is by no means easy, in many cases, to decide between *Brachyphyllum* and *Thuites* as the most fitting generic designation for coniferous twigs, and possibly some of the fossils referred to *Thuites expansus* are fragments of a *Brachyphyllum*.

Stonesfield? *Mantell Coll.*

V. 4676. A branched shoot with leaf-areas of the *Brachyphyllum* type.

Barnack (Inferior Oolite).

V. 4688. Similar to **11,130** (Pl. IX. Fig. 5), but larger and more obscurely preserved. The leaf-areas are rather broader and the apices of the leaves less prominent than in typical examples of *Thuites expansus.* *Egerton Coll.*

Other specimens:—**V. 3458** (Sevenhampton, *Buckman Coll.*); **V. 6586.**

Brachyphyllum, sp. β.

(Pl. XII. Figs. 9, 9*a*.)

In a paper *On some Undescribed Coniferous Fruits from the Secondary Rocks*, Mr. Carruthers refers in an appendix [1] to some branches from the Oxford Clay of Chippenham, Cambridgeshire, and Christian Malford, Wiltshire, and gives drawings of the specimens, but he wisely refrains from appending names to the fossils. A small piece of one of the specimens is shown in Pl. XII. Fig. 9. Without venturing to refer these coniferous fragments to a specific type, they may be briefly described as possibly twigs of a *Brachyphyllum.* The form of the polygonal leaf-areas agrees also with those on branches of Cretaceous conifers placed in the genus *Geinitzia*; but the question of a generic name is unimportant, as we are unable to refer the fossils with any degree of certainty to their family position among the Coniferæ. The vegetative twigs figured by Saporta [2] as *Brachyphyllum nepos* from the Kimeridgian and Corallian bear a close resemblance to the English Oxford Clay species.

52,837. Pl. XII. Figs. 9, 9*a*.

Figured also by Carruthers (Geol. Mag. vol. vi. pl. ii. fig. 12) and by Schimper [Trait. pal. vég. pl. lxxv. fig. 8 (copy of Carruthers' figure)]. A carbonised stem 20·5 cm. long and 1·5 cm. broad: the upper part is covered with carbonaceous polygonal areas, in the centre of each of which is a prominent rounded umbo; a fairly well-marked median keel or ridge and a similar transverse ridge extend across each polygonal area. In the lower

---

[1] Carruthers (69), p. 7, pl. ii. figs. 11–13.

[2] Saporta (84), p. 356, pls. clxviii.–clxx.

part of the stem the areas are more elongated and lozenge-shaped. The figures 9, *9a*, Pl. XII. illustrate the form of the polygonal areas.

Christian Malford, Wiltshire (Oxford Clay). *Cunnington Coll.*

40,547. Figured by Carruthers (Geol. Mag. vol. vi. pl. ii. fig. 13). An imperfect specimen of a branch bearing several short lateral branchlets, similar in habit to *Brachyphyllum.*

Christian Malford (Oxford Clay). *Bowerbank Coll.*

21,436. Figured by Carruthers (loc. cit. pl. ii. fig. 11). A piece of a carbonised branch 16 cm. long; details very obscure. Possibly the same species as 52,837 (Pl. XII. Fig. 9), but in a decorticated condition.

Chippenham, Cambridgeshire (Oxford Clay). *W. Buy Coll.*

Other specimens:—**21,448** (*W. Buy Coll.*); **36,773**; **40,548** (*Bowerbank Coll.*).

Genus **CONITES**, Sternberg.

[Flor. Vorwelt, iii. p. 36, 1823.]

I have elsewhere[1] discussed the advisability of making use of this designation for cones of doubtful affinity. In his recently published monograph on the Flora of Tonkin, Professor Zeiller[2] has also used this generic name for male and female gymnospermous flowers of uncertain position.

Conites primævus (Lindley & Hutton).

[Foss. Flor. pl. cxxxv. 1834.]

1834. *Pinus primæva*, Lindley & Hutton, Foss. Flor. pl. cxxxv.
1845. *Pitys (Pinus) primæva*, Unger, Synopsis, p. 198.
1848. *Pinus primæva*, Bronn, Ind. Pal. p. 982.
1850. *Pinites primæva*, Göppert, Foss. Conif. p. 222.
P. primævus, Unger, Gen. spec. foss. plant. p. 360.
1866. *P. primævus*, Carruthers, Geol. Mag. vol. iii. p. 545.
1867. *Cycadeostrobus primævus*, ibid. vol. iv. p. 105.
1872. *Zamiostrobus primævus*, Schimper, Trait. pal. vég. vol. ii. p. 202.
1885. *Z. primævus*, Zigno, Flor. foss. Oolit. vol. ii. p. 152.
1894. *Z. primævus*, Woodward, Lower Ool. p. 599.

---

[1] Seward (95), p. 113.
[2] Zeiller (03), p. 212.

Lindley & Hutton founded this species on a single cone, 2·5 cm. long and 1·6 cm. broad, obtained from rocks at Burcott Wood, near Towcester, said to be of Inferior Oolite age. I have not seen the type-fossil, but the specimen as represented in the drawings does not afford sufficient evidence to justify its reference to such a genus as *Pinites*. We can only leave the fossil as a form of *Conites* of uncertain systematic position; there appear to be no good reasons for regarding it as Cycadean.

Conites depressus (Carruthers).

[Geol. Mag. vol. vi. p. 2, pl. ii. fig. 10, 1869.]

1869. *Pinites depressus*, Carruthers, Geol. Mag. vol. vi. p. 2, pl. ii. fig. 10.
1875. *P. dejectus*, Blake, Quart. Journ. Geol. Soc. vol. xxxi. p. 222.
1884. *P. dejectus*, Damon, Geol. Weymouth, p. 68.
1895. *P. dejectus*, Woodward, Lower Ool. p. 401.

The original specimen figured by Carruthers is in the British Museum Collection (V. 6370).

The species was thus diagnosed in 1869:—

"Cone small, cylindrical, depressed at the apex; scales short, very broad, thin at the apex."

Carruthers describes the specimen as imperfect, and probably a young cone; it was obtained from the Kimeridge Clay of Weymouth.

V. 6370. Figured by Carruthers (Geol. Mag. vol. vi. pl. ii. fig. 10, 1869). The specimen has partially fallen to pieces, and is much too imperfect to refer to a definite family or generic position.

Weymouth (Kimeridge Clay).

CONIFEROUS WOOD.

Species of coniferous wood, too imperfectly preserved to examine microscopically, are recorded from the Stonesfield Slate, the Oxford Clay near Peterborough, the Coral Rag of Wotton, and elsewhere.

V. 2599. Wood partially converted into lignite.

Fletton, near Peterborough (Oxford Clay).

Presented by A. N. Leeds, Esq., 1890.

40,540. Fragment of lignite (jet?).

Weymouth (Kimeridge Clay). *Bowerbank Coll.*

CONIFERALES ?

? NAGEIOPSIS, sp.

[Pecopteris diversa, Phillips.]

[Geol. Oxford, p. 168, Diag. xxviii. fig. 1.]

In Phillips' *Geology of Oxford* a specimen from Stonesfield, now in the Oxford Museum, is figured as *Pecopteris diversa.* An examination of the fossil leads me to regard it as an imperfectly preserved coniferous shoot, possibly an example of the genus *Nageiopsis.* This genus has been employed by Fontaine [1] as a name for certain vegetative shoots from the Potomac beds of North America which bear a resemblance to those of recent species of the genus *Podocarpus.* The specimens in the Oxford Museum are, however, too obscure to determine; it is probable they belong to some member of the Coniferæ, possibly to *Nageiopsis*, and cannot be retained under Phillips' designation.

41,386. An obscure fragment which may be specifically identical with the larger specimen figured by Phillips. *Morris Coll.*

LEAVES OF DOUBTFUL SYSTEMATIC POSITION.

Phyllites, sp.

(Pl. XI. Figs. 5–6.)

The two drawings reproduced in Pl. XI. Figs. 5–6 represent two impressions of a leaf from Stonesfield. The fossil occurs as a brown stain on a fine-grained sandstone; the impression shown in Fig. 5 is that of a leaf the lamina of which is 3·7 cm. long and 2·5 cm. broad, with a portion of petiole. The preservation is not sufficiently good to afford any indication of the finer veins, but three main veins are clearly shown in both Figs. 5 and 6; the vascular tissue branches at the summit of the leaf-stalk into three approximately equal ribs, and from the rib shown on the right-hand side of the lamina in Fig. 6 two short branches are given off towards the edge of the leaf.

---

[1] Fontaine (89); see also Seward (00), p. 288.

Had the specimen been found in rocks known to contain the remains of Angiosperms, there would be no hesitation in identifying it as the leaf of a Dicotyledon; but seeing that we know of no undoubted Angiospermous fossil in Jurassic strata, it is of the utmost importance to demand satisfactory evidence before identifying a plant, or fragment of a plant, as an Angiosperm.

The statement made by Professor Bessey that "Monocotyledons and Dicotyledons probably appeared at about the same time early in the Mesozoic or late in the Palæozoic"[1] is founded on specimens that have been referred to the Angiosperms on evidence that is entirely without value. It has been stated by Deane[2] that well-developed Dicotyledons occur in the Oxley beds of Australia, which are regarded as Jurassic; but I have not seen any figures in support of this announcement, and in the absence of more data we cannot attach much importance to the statement. Two imperfectly preserved fragments of leaves have been figured by Kurtz from the Lias of South America[3] as doubtful fossils resembling Dicotyledons; they may perhaps represent portions of a *Dictyophyllum* frond; they are at all events valueless as evidence in support of the existence of Liassic Angiosperms.

There is no doubt that some of the supposed earliest Dicotyledons described from Portuguese rocks[4] are fronds of ferns similar in form and venation to the orbicular 'mantel-leafs' of the recent epiphytic fern *Platycerium*: in the present case it is difficult to believe that we have to do with the leaf of a fern. The genus *Gnetum*—the sole living representative of the Gnetales, a section of Gymnosperms—is characterised by leaves of the Dicotyledonous type, and there is always the possibility to be borne in mind that some of the so-called fossil Dicotyledons may be the leaves of *Gnetum*. So far as it is possible to base an opinion on the imperfect specimen before us, the venation characters do not conform closely to those of the leaves of recent species of *Gnetum*.

As regards the geological age of the fossil, the impression shown in Fig. 6, Pl. XI. is entered in the Museum Register as

---

[1] Bessey (97), p. 11.
[2] Deane (97), p. 858.
[3] Kurtz (01), pl. iii. figs. 10–11.
[4] Saporta (94); Ward (95), (96).

having been obtained from Stonesfield, a statement which there is no reason to suppose inaccurate. Moreover, the rock appears to be identical with that in which the majority of the Stonesfield plants are found. The available evidence seems to me to point to the identification of this imperfect leaf as a Dicotyledon from rocks of the Great Oolite Series. No trace of an Angiospermous plant has been recognised in the rich Inferior Oolite flora of East Yorkshire, nor has the Wealden flora, as represented by the plant-beds in the Sussex cliffs, afforded any evidence whatever pointing to the occurrence of Dicotyledonous or Monocotyledonous plants. These facts must of necessity make one very cautious in accepting this somewhat obscure and isolated specimen as proof of the existence of Dicotyledons in the vegetation which has left fragmentary relics in the Stonesfield beds.

Assuming that the Stonesfield leaf is that of a Dicotyledonous plant, it is impossible to determine its precise position. The venation characters, so far as they are visible, may be closely matched in the leaves of more than one family of Dicotyledons. I am disposed to think that the best course to follow is to assign the fossil, with some hesitation, to the convenient genus *Phyllites*, using the name as a designation for Dicotyledonous leaves which cannot be referred with any certainty to a particular family.

I do not regard the evidence furnished by the Stonesfield leaf as sufficiently satisfactory to be regarded as furnishing proof of the existence of Dicotyledons during the Jurassic era, but it suggests the possibility that the highest class of plants had made its appearance long before it assumed the dominant position to which it was suddenly raised in the Lower Cretaceous period. Charles Darwin wrote to Hooker in 1879—"The rapid development, as far as we can judge, of all the higher plants within recent geological times, is an abominable mystery."[1] On another occasion he wrote—"Nothing is more extraordinary in the history of the vegetable kingdom, as it seems to me, than the apparently very sudden or abrupt development of the higher plants. I have sometimes speculated whether there did not exist somewhere during long ages an extremely isolated continent, perhaps near the South Pole."[2]

---

[1] Darwin (03), ii. p. 20.
[2] Darwin (87), vol. iii. p. 248.

It is highly probable that the suddenness with which the Dicotyledons took their place in the vegetation of the world is exaggerated by the scantiness of our fossil records. We should expect to obtain evidence pointing to the gradual increase in the number of Angiosperms as we investigate the plant-bearing strata below those in which these plants are richly represented. There is, in short, no *a priori* improbability that Dicotyledons existed ages before they attained to a position of overwhelming importance, and indeed it is highly probable that this was the case.

I have recently ventured to call the attention of botanists to the importance of undertaking a thoroughly critical revision of the records of fossil Angiosperms.[1] Species quoted by writers as evidence bearing on questions connected with the phylogeny of flowering plants have in many cases been determined by authors whose lack of botanical knowledge renders their records of doubtful value, if not positively misleading and pernicious. The custom of describing some of the oldest known Angiosperms as Proangiosperms and generalised types is likely to mislead those who are not familiar with the fragmentary and doubtful nature of the records. We require an organised exploration of the later plant-bearing beds and of the wealth of material already collected, which should be taken in hand by experienced palæobotanists in conjunction with botanists who possess wide and accurate knowledge of recent Angiosperms.

SUPPOSED MONOCOTYLEDONOUS PLANT.

Aroides Stutterdi, Carruthers.

[Geol. Mag. vol. iv. p. 146, pl. viii. fig. 2, 1867.]

1867. *Aroides Stutterdi*, Carruthers, Geol. Mag. vol. iv. pl. viii. fig. 2.
1871. *A. Stutterdi*, Phillips, Geol. Oxford, p. 168, Diag. xxxii. figs. 12–13.
1873. *A. Stutterdi*, Zigno, Flor. foss. Oolit. vol. ii. p. 2.
A. Stutterdi, Sharp, Quart. Journ. Geol. Soc. vol. xxix. p. 295.
1875. *A. Stutterdi*, Judd, Geol. Rutland, p. 165.
1886. *A. Stutterdi*, Gardner, Geol. Mag. vol. iii. p. 200.
1894. *A. Stutterdi*, Woodward, Lower Ool. p. 195.
1896. *A. Stutterdi*, Seward, Annals Bot. vol. x. p. 212.

---

[1] New Phytologist, December, 1903.

In 1867 Carruthers described some specimens from the Stonesfield Slate under the name *Aroides Stutterdi*, which he regarded as fragments of an Aroideous spadix comparable with that of the recent *Xanthosoma*. It was subsequently suggested that the fossil might represent a portion of an anal sac of a Crinoid,[1] but this view is not shared by Dr. Bather, of the British Museum, who examined some specimens of Carruthers' fossil.[2] The Stonesfield fossils have the form of cylindrical casts covered by peltate plates — subquadrangular or hexagonal in outline—dovetailing into one another by irregularly dentate margins. I am unable to offer any opinion as to the nature of the specimens, beyond stating that in my opinion they do not afford any satisfactory evidence of the occurrence of Monocotyledonous plants in the Stonesfield Slate.

In addition to the specimens in the British Museum others may be seen in the Oxford Museum and in the Jermyn Street collection.

V. 3442. A specimen showing four rows of hexagonal projecting areas arranged in vertical series.
Stonesfield. *Brodie Coll.*

V. 5585. A very good specimen showing the calcareous plates more clearly than in the figures published in Carruthers' paper.
Presented by Mr. S. Stutterd.

52,871. In this example the plates are less regular in shape than in the preceding example. *S. Sharp Coll.*

---

[1] Quoted by Gardner (86), p. 198.
[2] Seward (96), p. 212.

CONCLUSION.

The following lists include the plants described in this volume arranged according to their geological age:—

I. TRIAS (Keuper).

Fragments of coniferous twigs (*Voltzia?*) and other fossils too imperfect to determine.

Localities.—Rowington (Warwickshire); Pendock (Worcestershire); Leicester.

II. RHÆTIC.

Equisetales.

Equisetites Muensteri, Sternb.

Lycopodiales.

Lycopodites lanceolatus (Brodie).

Filicales.

Clathropteris platyphylla (Göpp.).

Gymnospermæ.

Carpolithes, sp.

Plantæ Incertæ Sedis.

? *Araucarites*, sp.

Indeterminable Fragments.

Localities. — Strensham (Worcestershire); neighbourhood of Bristol; Binton (Warwickshire).

III. LIAS.

Thallophyta.

Indeterminable specimens formerly referred to the algæ.

EQUISETALES.

? *Equisetites Muensteri*, Sternb.

FILICALES.

Thinnfeldia rhomboidalis, Ett. (Kurr).
Ctenopteris cycadea, Brongn.

CYCADOPHYTA.

Cycadites rectangularis, Brauns.
? *C. rectangularis*, Brauns.
Otozamites obtusus (L. & H.).
Cycadeoidea gracilis (Carr.).
? *C. pygmæa*, L. & H.

CONIFERALES.

Pagiophyllum peregrinum (L. & H.).
Coniferous wood:
    a. Casts.
    b. Petrifactions.
Araucarioxylon Lindleii (Witham).
? *A. Lindleii* (Witham).
Cupressinoxylon Barberi, sp. nov.
Jet.

Localities. — Lyme Regis (Dorsetshire); Charmouth (Dorsetshire); Whitby and Staithes (Yorkshire); Northampton, Litchborough, Kilsby, Bugbrook, Roade, Braunston (Northamptonshire); Grantham (Lincolnshire).

IV. OOLITE.

A. Inferior Oolite.

THALLOPHYTA (?).

Girvanella pisolitica, Weth.

FILICALES.

Laccopteris Woodwardi (Leck.).
Dictyophyllum, sp.

CYCADOPHYTA.

Williamsonia Bucklandi (Ung.).
W. pecten (Phill.).
Otozamites, sp. [Cf. *O. obtusus* (L. & H.).]

CONIFERALES.

Brachyphyllum, sp. *a*.
Araucarites sphærocarpus, Carr.
A. Cleminshawi, Mans.-Pleyd.

Localities.—Cleeve Hill (Gloucestershire); Stamford, Ponton, and Collyweston (Lincolnshire); Wansford, Rothwell, Barnack, and Weekley (Northamptonshire); Sherborne and Charmouth (Dorsetshire); Bruton (Somersetshire).

B. Great Oolite.

THALLOPHYTA.

? *Algites furcatus* (Brongn.).
Solenopora jurassica, Brown.

FILICALES.

Laccopteris Woodwardi (Leck.).
Tæniopteris vittata, Brongn.
Sphenopteris, sp. *a*.
S., sp. *b*.
Sagenopteris Phillipsi, Brongn.
Thinnfeldia speciosa, Ett.

GINKGOALES.

Ginkgo digitata (Brongn.).
Baiera Phillipsi, Nath.

CYCADOPHYTA.

Williamsonia pecten (Phill.).
Cycadeoidea squamosa (Brongn.).
Zamites megaphyllus (Phill.).
Ctenis latifolia (Brongn.).
Sphenozamites Belli, sp. nov.
Podozamites stonesfieldensis, sp. nov.
Carpolithes diospyriformis, Sternb.

CONIFERALES.

Araucarites ooliticus (Carr.).
A. Brodiei, Carr.
A. sp.
? *Nageiopsis*, sp.
Thuites expansus, Sternb.
Brachyphyllum, sp. *a.*

? ANGIOSPERMÆ.

Phyllites, sp.

Localities.—Stonesfield (Oxfordshire); Sevenhampton and Eyeford (Wiltshire); Chedworth (Gloucestershire); Kingsthorpe and Leckhampton (Northamptonshire).

C. Oxfordian.

GYMNOSPERMÆ.

Carpolithes, sp.

CONIFERALES.

Araucarites sphæricus (Carr.).
Thuites, sp.
Brachyphyllum, sp. *β.*
Conites primævus (L. & H.).
Coniferous wood.

Localities.—Scarborough (Yorkshire); Christian Malford (Wiltshire); Chippenham (Cambridgeshire); Peterborough (Northamptonshire).

D. Corallian.

GYMNOSPERMÆ.

Carpolithes conicus, L. & H.
C. sp.

CONIFERALES.

Araucarites ooliticus (Carr.).

Localities.—Malton (Yorkshire).

E. Kimeridgian.

[*Caulerpa Carruthersi*, Murray.]

CONIFERALES.

Conites depressus (Carr.).
Coniferous wood.

Localities.—Weymouth, Sandsfoot (Dorsetshire).

It is of interest to notice the geographical range of the species of Rhætic and Jurassic (Liassic and Oolitic) plants represented in English floras. In attempting to illustrate the distribution in space of fossil species one is confronted with the difficulty of deciding to what extent species recorded from Western Europe may be regarded as identical with similar forms described from distant regions. The resemblance of one flora to another is usually obscured by the use of different generic or specific names for plants, which are either identical or represent closely allied members of the same family. This diversity in nomenclature is, to some extent, the result of geographical separation; an author naturally hesitates to assign the same specific name to plants from India and Europe unless the evidence as to identity is convincing. On the other hand, wide separation in space has often been allowed to exercise a misleading influence in the determination of species. Another reason for the use of different names for plants which are either specifically identical or very closely allied, is to be found in the individual preferences of authors in the choice of possible generic designations for a particular type.

In order to obtain a clear idea of the botanical relationship of one flora to another it is essential to devise some method by which distinctions, either imaginary or exaggerated, between plants recorded by different writers under distinct names may be eliminated.

As Darwin wrote in the *Origin of Species*,[1] "It is notorious on what excessively slight differences many palæontologists have founded their species; and they do this the more readily if the specimens come from different sub-stages of the same formation."

Our aim is to ascertain to what extent homotaxial floras in different parts of the world were characterised by similar botanical facies. The subject of the geographical distribution of plants during past ages is one which has not received sufficient attention; it is true our data are too meagre to admit of sweeping generalisations, but we have enough evidence at our disposal to enable us to form an estimate of the comparative composition of widely separated floras. If we confine ourselves to a comparison of published lists

[1] Darwin (00), p. 423.

of various local floras, we shall inevitably exaggerate the differences in their composition by adopting the nomenclature used by individual authors. We must seek to discover whether types recorded from one region are represented by similar types in other parts of the world; it is of less importance to endeavour to decide if a southern hemisphere plant should be regarded as specifically identical with a northern form than to discover in what degree the general facies of an European flora is repeated in a flora from southern latitudes. In the following list I have therefore adopted a liberal interpretation of specific designations, and have endeavoured to show to what extent representative or identical types occurred in various parts of the world during the Rhætic and Jurassic eras. In taking considerable liberties with the nomenclature of other authors, I do not necessarily mean to express disagreement with them as regards their interpretation of affinity; but my aim is to avoid the danger of allowing slight differences—whether of specific rank or not—to obscure the broad relationships of floras. The method of comparison is adopted primarily for the purpose of instituting a botanical comparison, rather than with the view of expressing an opinion as to the relative age or stratigraphical position of the rock.[1]

In the following table (pp. 164–167) I have indicated, in the case of the Oolitic species, the sub-stages from which the fossils have been obtained. I. O. = Inferior Oolite; G. O. = Great Oolite; Cor. = Corallian; Oxf. = Oxfordian; Kim. = Kimeridgian.

RHÆTIC.—The English Rhætic species constitute too meagre a sample of a Rhætic flora to render necessary any discussion as to the distribution and composition of the Rhætic vegetation. The wide range of *Equisetites Muensteri* and allied forms and of the fern *Clathropteris* may be taken as an index of the uniformity of Rhætic floras throughout the world. In addition to the countries named in the table Chili[2] and Honduras may be mentioned as two other regions from which Rhætic plants have been recorded. Newberry has drawn attention to the uniformity of the Rhætic

---

[1] Seward (03).

[2] Solms-Laubach (99).

vegetation in both the Old and New Worlds; he wrote in 1888: "We shall . . . . look now with eagerness to South America for the identification there of this Mesozoic flora, which we have found in full development in Virginia, New Mexico, Sonora, and now in Honduras. It has been recognised in Australia, New Zealand, India, Tonquin, China, Turkestan, and various parts of Europe. Hence with its discovery in South America we shall see it reaching as a girdle around the entire globe."[1] Before this was written Geinitz[2] had recorded Rhætic plants from the Argentine, and more recently Szajnocha, Solms-Laubach, and Kurtz have made further contributions towards the completion of this girdle of Rhætic floras. One of the richest floras of the Rhætic period is that of Tonkin, which has recently received exhaustive treatment by Professor Zeiller.[3] The assemblage of types affords another demonstration of the uniform character of the vegetation of this period; it is true that a few types which occupy a prominent position in the Rhætic floras of Scania and Franconia have not been recorded from Tonkin, but in its general composition the vegetation of the Far East exhibits a striking agreement with that of the northern hemisphere.

LIAS.—The vegetation which has left traces in rocks of Liassic age differs but little from that of the Rhætic period, and indeed it is often almost impossible to distinguish clearly between floras of the two periods.

The plant which I have referred to *Thinnfeldia rhomboidalis* is an example of a genus which was very widely spread in both the Rhætic and Liassic periods. A similar form, though probably distinct specifically, originally described by Morris[4] as *Pecopteris odontopteroides* (= *Thinnfeldia odontopteroides*), is an exceedingly abundant fossil in the Stormberg Series of Cape Colony, in beds of Rhætic age in Australia, and in South America. The two species of *Cycadeoidea*, *C. gracilis* and *C. pygmæa*, are in all probability not peculiar to England; but it is impossible to define specific

[1] Newberry (88).
[2] Geinitz (76).
[3] Zeiller (02), (03).
[4] Morris (45), p. 249, pl. vi. figs. 2–4.

types of Cycadean stems with any certainty when we have nothing more than imperfectly preserved external features to guide us.

The records given in the above list, which is far from complete, show the geographical range of the species described in this and in the preceding volume on the Jurassic floras of England, or of plants which, though not identical with the English types, may be fairly considered as representative species.

OOLITE.—The wide geographical range of the Oolitic flora is clearly demonstrated in the table. The occurrence of a few Jurassic plants in Madagascar has recently been noted by Professor Zeiller,[1] and in a paper in course of publication I have described several British Jurassic species in plant-bearing strata of Victoria.

It is hardly possible to discriminate between the various forms of vegetative shoot referred to the genus *Equisetites*, but there is no doubt as to the abundance and wide distribution of the Equisetaceæ during the Jurassic era. The Equisetales differ from some of the other representatives of Pteridophytes in having maintained their position through several ages as plants which are still capable of living in abundance under very different climatal conditions. Among the ferns represented in the Jurassic floras of Europe the Matonineæ and Dipteridinæ are of especial interest: *Matonidium* and *Laccopteris* demonstrate the wide range of the former and *Dictyophyllum* illustrates the abundance of the latter family during a considerable part of the Mesozoic epoch. The range of *Dictyophyllum* as given in the above table might be legitimately extended by the inclusion of *Clathropteris*, *Protorhipis*, and other members of the Dipteridinæ, which cannot in all cases be clearly distinguished from one another. The species *Klukia exilis* points to the former extension of the Schizæaceæ into regions where representatives of the family are no longer met with. *Todites Williamsoni*,[2] frequently found in a fertile condition, indicates an almost worldwide range of Osmundaceous ferns very closely allied to the recent *Todea barbara*, Hook., of South Africa, Eastern Australia, and New Zealand. The exceedingly common type

---

[1] Zeiller (00).
[2] Seward & Ford (03).

referred to under the comprehensive designation *Cladophlebis denticulata* is one of the most ubiquitous of Mesozoic ferns; but in this case the identification of the species is usually founded on sterile fronds which may well belong to several distinct families.

Among Gymnosperms the Ginkgoales and Cycadophyta are examples of classes of plants which reached their maximum development during the Mesozoic era. The numerous species of Conifers described from Jurassic rocks and referred to a large number of genera are in most cases much less satisfactory from a botanical point of view; we have in most instances to be satisfied with a purely artificial designation which affords no clue as to their position among recent families of the Coniferales. As already pointed out, the exceedingly abundant species, *Thuites expansus*, which is by far the commonest plant in the Stonesfield Slate, may be a representative of the Cupressineæ, but the form of the cones lends an element of doubt to its inclusion in that family. The most striking exceptions are the genera *Araucarites* and *Araucarioxylon*; the former name being usually applied to Cones and the latter to wood possessing the anatomical characters of recent Araucarieæ. As Sir William Thiselton-Dyer[1] pointed out in 1878, the genus *Araucaria* affords one among several indications of the great antiquity of our present southern flora; it has long been extinct in northern regions, but in the Mesozoic era members of the family are known to have existed in several parts of Europe, in North America, in Chili,[2] India, Australia, South Africa, and elsewhere.

The greater number of species included in the above list of Oolitic plants have been obtained from the Inferior Oolite plant beds of East Yorkshire and from the Great Oolite of Stonesfield. In several cases I am unable to recognise any differences between plants from the two horizons which can reasonably be regarded as indices of specific distinction. Peculiar species occur in both the Inferior Oolite and Great Oolite floras, but on the whole the general facies is very similar; among species from the Stonesfield Slate that do not appear to be represented in the Inferior Oolite vegetation of Yorkshire we have *Thinnfeldia speciosa*, *Cycadeoidea*

---

[1] Thiselton-Dyer (78).
[2] Solms-Laubach (99).

squamosa, *Zamites megaphyllus*, *Ctenis latifolia*, *Sphenozamites Belli*, *Podozamites stonesfieldensis*, and *Carpolithes diospyriformis*. Some of the cones of *Araucarites* met with in the Stonesfield Slate may also be peculiar types, but similar Araucarian cones occur in Inferior Oolite rocks.

The most striking conclusion forced upon us by a general survey of the Oolitic floras is the similarity in the composition of the vegetation during the Jurassic era throughout the greater part of the world. We cannot, I believe, deduce any evidence from such data as we possess in favour of the existence of well-defined botanical provinces during the Rhætic, Jurassic, or Wealden periods.

BIBLIOGRAPHY.

Agassiz, A. (92). Dredging Operations in the Eastern Pacific. *Nature*, vol. xlv. p. 280.

Andrae, K. T. (53). Fossile flora Siebenbürgens und des Banates. *Abh. k.-k. geol. Reichs.* vol. ii. Abth. 3, No. 4, 1853.

Balfour, J. H. (72). Palæontological Botany. Edinburgh, 1872.

Barber, C. A. (98). *Cupressinoxylon vectense*; a fossil Conifer from the Lower Greensand of Shanklin, in the Isle of Wight. *Annals Bot.* vol. xii. p. 329.

Barrow, G. (82). See *Fox-Strangways & Barrow.*

——— (85). See *Fox-Strangways, Reid, & Barrow.*

Bartholin, C. T. (92). Nogle i den bornholmske Juraformation forekommende Plante forsteninger. *Bot. Tid. Bot. For. Kjövenhavn*, vol. xviii. Heft 1, p. 12, 1892.

——— (94). Ibid. vol. xix. p. 87.

Bayer, E. (01). See *Frič & E. Bayer.*

Beche, H. T. de la (24). Remarks on the Geology of the South Coast of England, from Bridport Harbour, Dorset, to Babbacombe Bay, Devon. *Trans. Geol. Soc.* vol. i. [2], p. 40.

——— (29). On the Lias of the Coast in the vicinity of Lyme Regis, Dorset. *Trans. Geol. Soc.* [2], vol. ii. p. 21.

——— (36). See *Buckland & de la Beche.*

Benecke, F. W. (68). Geognostisch. Paläontologische Beiträge, vols. i.–ii. Munich, 1868–76.

——— (98). *Diplopora* und einige andere Versteinerungen im Elsass-lothringischen Muschelkalk. *Mitt. Geol. Landes. von Elsass-Lothringen*, iv. p. 277.

Berger, H. A. C. (32). Die Versteinerungen der Fische und Pflanzen im Sandsteine der Coburger Gegend. Coburg, 1832.

Berry, E. W. (03). The American Species referred to *Thinnfeldia. Bull. Torrey Bot. Club*, vol. xxx. p. 438.

Bessey, C. E. (97). The Phylogeny and Taxonomy of Angiosperms. *Bot. Gaz.* vol. xxiv.

Bird, J. (22). See *Young & Bird.*

Blake, J. F. (75). On the Kimeridge Clay of England. *Quart. Journ. Geol. Soc.* vol. xxxi. p. 196.

——— (76). See *Tate & Blake.*

——— and *W. H. Hudleston* (77). On the Corallian Rocks of England. *Quart. Journ. Geol. Soc.* vol. xxxiii. p. 260.

Blanckenhorn, M. (86). Die fossile Flora des Buntsandsteins und des Muschelkalks der Umgegend von Commern. *Palæontographica*, vol. xxxii. p. 117.

Blanford, W. T. (84). On the Classification of Sedimentary Strata. *Geol. Mag.* vol. i. p. 318.

——— (93). See *Medlicott & Blanford.*

Bornemann, J. G. (56). Ueber organische Reste der Lettenkohlengruppe Thüringens. Leipzig, 1856.

Boscawen, L. L., Ibbetson, and *J. Morris* (48). Notice on the Geology of the neighbourhood of Stamford and Peterborough. *Brit. Assoc. Rep.* 1848 (Oxford, 1847), p. 127.

Bower, J. A. (73). Whitby Jet and its Manufacture. *Journ. Soc. Arts*, vol. xxii. p. 80, 1874.

Brauns, D. (62). Der Sandstein bei Seinstedt unweit des Fallsteins und die in ihm vorkommenden Pflanzenreste. *Palæontographica*, vol. ix. p. 47.

Brodie, P. B. (42). On the discovery of Insects in the Lower Beds of the Lias of Gloucestershire. *Brit. Assoc. Rep.* 1842, p. 58.

——— (45). A History of Fossil Insects in the Secondary Rocks of England. London, 1845.

——— (87). Notes on the Upper Keuper Section at Shrewley and on the Trias generally in Warwickshire. *Quart. Journ. Geol. Soc.* vol. xliii. p. 540.

——— and *J. Buckman* (45). On the Stonesfield Slate of the Cotteswold Hills. *Quart. Journ. Geol. Soc.* vol. i. p. 220.

Brongniart, A. (22). Sur la classification et la distribution de végétaux fossiles en général, et sur ceux des Terrains de Sédiment Supérieur en particulier. *Mém. Mus. d'Hist. Nat.* viii. p. 297.

——— (23). Observations sur les Fucoides et sur quelques autres plantes marine fossiles. *Mém. Hist. Nat. Soc.* vol. i. p. 301.

——— (25). Note sur les végétaux fossiles de l'Oolite à fougères de Mamers. *Ann. Sci. Nat.* vol. iv. p. 417.

——— (28). Prodrome d'une histoire des végétaux fossiles. Paris, 1828.

——— (28^2). Histoire des végétaux fossiles. Paris, 1828.

——— (49). Tableau des genres de végétaux fossiles. (Extrait du Dictionnaire d'histoire naturelle, vol. xiii.) Paris, 1849.

Bronn, H. G. (48). Index Palæontologicus, Nomenclator. Stuttgart, 1848.

Brown, A. (94). On the Structure and Affinities of the genus *Solenopora*. *Geol. Mag.* [4], vol. i. p. 145, 1894.

Buckland, W. (24). Notice on the *Megalosaurus*, or Great Fossil Lizard, of Stonesfield. *Trans. Geol. Soc.* vol. i. [2], p. 390.

——— (27). Note on Cycadeoideæ. *Proc. Geol. Soc.* vol. i. p. 80.

——— (28). On the Cycadeoideæ, a family of fossil plants found in the Oolite Quarries of the Isle of Portland. *Geol. Trans.* vol. ii. [2], p. 395.

——— (37). Geology and Mineralogy considered with reference to Natural Theology. London, 1836.

——— and *H. T. de la Beche* (36). On the Geology of the neighbourhood of Weymouth and the adjacent parts of the Coast of Dorset. *Trans. Geol. Soc.* vol. iv. [2], p. 1.

Buckman, J. (45). See *Murchison, Buckman, & Strickland.*

——— (45). See *Brodie & Buckman.*

——— (50). On some Fossil Plants from the Lower Lias. *Quart. Journ. Geol. Soc.* vol. vi. p. 413.

——— (63). *Stricklandia acuminata.* [Anonymous.] *The Geologist,* vol. vi. p. 395.

Bunbury, C. J. F. (51). On some Fossil Plants from the Jurassic Strata of the Yorkshire Coast. *Quart. Journ. Geol. Soc.* vol. vii. p. 179.

Capellini, G. and *Graf zu Solms-Laubach* (91). I Tronchi di Bennettitee dei musei Italiani. *Mem. R. Accad. Sci. inst. Bologna* (5), vol. ii. p. 161, 1891.

Carruthers, W. (66). On some Fossil Coniferous Fruits. *Geol. Mag.* vol. iii. p. 534.

——— (67). On some Cycadean Fruits from the Secondary Rocks of Britain. *Geol. Mag.* vol. iv. [1], p. 101.

——— (67²). On an Aroideous Fruit from the Stonesfield Slate. *Geol. Mag.* vol. iv. p. 146.

——— (68). British Fossil Pandanaceæ. *Geol. Mag.* vol. v. p. 153.

——— (69). On some Undescribed Coniferous Fruits from the Secondary Rocks of Britain. *Geol. Mag.* vol. vi. [1], p. 1.

——— (70). On Fossil Cycadean Stems from the Secondary Rocks of Britain. *Trans. Linn. Soc.* vol. xxvi. p. 675.

——— (71). On two Undescribed Coniferous Fruits from the Secondary Rocks of Britain. *Geol. Mag.* vol. viii. p. 540.

——— (71²). On some supposed Vegetable Fossils. *Quart. Journ. Geol. Soc.* vol. xxvii. p. 443.

Carruthers, W. (72). Notes on Fossil Plants from Queensland, Australia. *Quart. Journ. Geol. Soc.* vol. xxviii. p. 1.

—— (77). Description of a New Species of *Araucarites* from the Coralline Oolite of Malton. Appendix to a paper by Blake & Hudleston, "On the Corallian Rocks of England." *Quart. Journ. Geol. Soc.* vol. xxxiii. p. 402.

Compter, G. (74). Ein Beitrag zur fossilen Keuperflora. *Nov. Act. Ac. Cæs. Leop.-Car.* vol. xxxvii. No. 3.

Conwentz, H. (90). Monographie der baltischen Bernsteinbäume. Danzig, 1890.

Coulter, J. M. and *C. J. Chamberlain* (01). Morphology of Spermatophytes. New York, 1901.

Dale, E. (00). See *Seward & Dale.*

Damon, R. (84). The Geology of Weymouth, Portland, and the Coast of Dorsetshire. Weymouth, 1884.

—— (88). A Supplement to the Geology of Weymouth. Weymouth, 1888.

Darwin, C. (87). Life and Letters of Charles Darwin. London, 1887.

—— (00). The Origin of Species. London, 1900.

—— (03). More Letters of Charles Darwin. Edited by F. Darwin and A. C. Seward. London, 1903.

Deane, H. (97). President's Address. *Proc. Linn. Soc. N.S. Wales,* part iv.

Engler, A. and *K. Prantl* (97). Die natürlichen Pflanzenfamilien. Nachtrag zu Teile ii.–iv. p. 19.

Etheridge, R. (82). Presidential Address. *Quart. Journ. Geol. Soc.* vol. xxxviii. Proc. p. 59.

Etheridge, R., jun. (80). See *Nicholson & Etheridge.*

Ettingshausen, C. von (52). Begründung einiger neuen oder nicht genau bekannten Arten der Lias und der Oolithflora. *Abh. k.-k. geol. Reichs.* vol. i. Abth. 3.

Feistmantel, O. (76). Fossil Flora of the Gondwana System. Jurassic (Oolitic) Flora of Kach. *Mem. Geol. Surv. India,* vol. ii. pt. 1.

—— (77). Ibid. Jurassic (Liassic) Flora of the Rajmahal Group in the Rajmahal Hills.

—— (77[2]). Ibid. Flora of the Jabalpur Group.

Fliche, P. (00). Contribution à la flore fossile de la Haute-Marne. *Bull. Soc. Sci. Nancy,* 1900.

—— (01). Un nouveau *Cycadeoidea. Bull. Soc. géol. France,* vol. i. [iv.], p. 193.

—— (03). Sur les Lycopodinées du Trias en Lorraine. *Compt. Rend.,* April 6, 1903.

Fontaine, W. M. (83). Contributions to the knowledge of the Older Mesozoic Flora of Virginia. *U.S. Geol. Surv. Mon.* vol. vii.

——— (89). The Potomac or Younger Mesozoic Flora. Ibid. vol. xv.

Fox-Strangways, C. (92). The Jurassic Rocks of Britain. Vol. i: Yorkshire. *Mem. Geol. Surv.*

——— and *G. Barrow* (82). The Geology of the Country between Whitby and Scarborough. *Mem. Geol. Surv., Quarter-Sheet* 95 *N.W.*

——— *C. Reid*, and *G. Barrow* (85). The Geology of Eskdale, Rosedale, etc. *Mem. Geol. Surv.* 1885.

Frič, A. and *E. Bayer* (01). Studien im Gebiete der Böhmischen Kreideformation. *Archiv. Naturwiss. Landes. Böhmen.* vol. xi.

Gardner, J. S. (86). On Mesozoic Angiosperms. *Geol. Mag.* vol. iii. [3], p. 193.

Geikie, A. (93). A Text-book of Geology (3rd ed.). London, 1893.

Geinitz, H. B. (76). Über rhaetische Pflanzen und Thierreste in den argentinischen Provinzen La Rioja, San Juan, und Mendoza. *Palæontographica*, Suppl. iii.

Geyler, H. T. (77). Über fossile Pflanzen aus der Juraformation Japans. *Palæontographica*, vol. xxiv. p. 221.

Göppert, H. R. (36). Die fossilen Farrnkräuter. *Nova Acta Ac. Cæs. Leop.-Car.* vol. xvii. (Suppl.).

——— (41). Die Gattungen der fossilen Pflanzen. *Bonn*, 1841.

——— (43). Ueber die fossilen Cycadeen überhaupt, mit Rücksicht auf die in Schlesien vorkommender Arten. *Ueberícht der Arbeiten und Veränderungen der Schlesischen Gesell. für Vart. Kultur.* 1843, p. 114. Breslau, 1844.

——— (50). Monographie der fossilen Coniferen. *Nat. Verhand. Holland. Maatschaf. Haarlem. Leiden*, 1850.

Gowan, J. (00). See *Seward & Gowan.*

Harris, G. F. (96). On the Analysis of Oolitic Structure. *Proc. Geol. Assoc.* vol. xiv. p. 59.

Hartz, N. (96). Planteforsteninger fra Cap. Stewart i Ostgrönland. *Meddelelser Grönl.* vol. xix. p. 217.

Heer, O. (76). Flora fossilis Helvetiæ. Zürich, 1876.

——— (81). Contributions à la flore fossile du Portugal. *Secc. Trab. Geol. Portugal.*

Holmes, T. V. (81). The Permian, Triassic, and Liassic Rocks of the Carlisle Basin. *Quart. Journ. Geol. Soc.* vol. xxxvii. p. 286.

——— (99). The Geology of the Country around Carlisle. *Mem. Geol. Survey.*

Hooker, J. D. (63). On *Welwitschia*, a new genus of Gnetaceæ. *Trans. Linn. Soc.* vol. xxiv.

Horton, W. S. (60). On the Geology of the Stonesfield Slate and its associate formations. *The Geologist*, p. 249, 1860.

Hudleston, W. H. (77). See *Blake & Hudleston.*

——— (78). The Yorkshire Oolites. *Proc. Geol. Assoc.* vol. v. p. 407.

Hull, E. (57). The Geology of the Country around Cheltenham. *Mem. Geol. Surv.*

Hutton, W. (31–37). See *Lindley & Hutton.*

Ibbetson, — (48). See *Boscawen, Ibbetson, & Morris.*

Jaeger, G. F. (27). Über die Pflanzenversteinerungen welche in dem Bausandstein von Stuttgart vorkommen.

Judd, J. W. (75). The Geology of Rutland. *Mem. Geol. Surv.*

Knowlton, F. H. (89). Fossil Wood and Lignites of the Potomac Formation. *Bull. U.S. Geol. Surv.* No. 56.

——— (90). A Revision of the genus *Araucarioxylon* of Kraus. *Proc. U.S. Nat. Mus.* vol. xii. p. 601.

——— (97). Appendix to the Geology of the Richmond Basin, Virginia, by *N. S. Shaler* and *J. B. Woodworth. Ann. Rep. U.S. Geol. Surv.* xix.

Krasser, F. (91). Über die fossile Flora der rhätischen Schichten Persiens. *Sitzb. k. Akad. Wiss. Wien.* vol. c. Abth. 1.

——— (00). Die von W. A. Obtrutshen in China und Centralasien 1893–1894 gesammelten fossilen Pflanzen. *Denksch. k. Akad. Wiss. Wien.* vol. lxx.

Kurr, J. G. (45). Beiträge zur fossilen Flora der Juraformation Württembergs. Stuttgart, 1845.

Kurtz, F. (01). Le Lias de la Piedra Pintada. *Revist. Mus. de la Plata*, vol. x. p. 225.

Lapparent, A. de (00). Traité de Géologie. Paris, 1900.

Leckenby, J. (64). On the Sandstones and Shales of the Oolites of Scarborough, with descriptions of some new Species of Fossil Plants. *Quart. Journ. Geol. Soc.* vol. xx. p. 74.

Lignier, O. (94). Végétaux fossiles de Normandie. Structure et affinités du *Bennettites Morierei*, Sap. & Mor. *Mém. Soc. Linn. Normandie*, vol. xviii.

——— (95). Végétaux fossiles de Normandie. *Mém. Soc. Linn. Normandie*, vol. xxviii. p. 5.

——— (03). Le fruit der *Williamsonia gigas* et les Bennettitales. *Mém. Soc. Linn. Normandie*, vol. xxi. p. 19.

Lindley, J. and *W. Hutton* (31–37). The Fossil Flora of Great Britain. *London*, 1831–37.

Lorenzo, G. de (97). Fossili del Trias medio di Lagonegro. *Palæontographica Italica*, vol. ii. p. 113, 1896. Pisa, 1897.

Lyell, C. (78). The Student's Elements of Geology (3rd ed.). London, 1878.

Mansel-Pleydell, J. C. (85). Notes on a Cone from the Inferior Oolite Beds of Sherborne. *Proc. Dorset Nat. Hist. Antiq. Field Club*, vol. v. p. 141.

Medlicott, H. B. and *W. T. Blanford* (93). Manual of the Geology of India (2nd ed.). Calcutta, 1893.

Miquel, F. A. W. (42). Monographia Cycadearum. Utrecht, 1842.

——— (47). Collectanea nova ad Cycadearum cognitionem. *Linnæa*, vol. xix. p. 411.

——— (51). Over de Rangschikking der fossiele Cycadeæ. [*Tijdsch. Wis. Nat. Wet.* vol. iv. p. 205, 1851.]

Möller, H. (02). Bidrag till Bornholms Fossila Flora. Part i. *Lunds Univ. Årsskrift*, vol. xxxviii.

——— (03). Ibid. Part ii. *K. Svensk. Vet.-Akad. Hand.* vol. xxxvi.

Moore, C. (67). On Abnormal Conditions of Secondary Deposits, etc. *Quart. Journ. Geol. Soc.* vol. xxiii. p. 449.

Morière, J. (80). Considérations générales sur la flore fossile et spécialement sur celle du Lias. *Bull. Soc. Linn. Normandie*, vol. iv. [3], p. 361.

Morris, J. (41). Remarks upon the Recent and Fossil Cycadaceæ. *Ann. & Mag. Nat. Hist.* vol. vii. p. 110.

——— (45). In Count Strzelecki's "Physical Description of New South Wales, etc.," p. 245. London, 1845.

——— (48). See *Boscawen, Ibbetson, & Morris.*

——— (53). On some Sections in the Oolitic District of Lincolnshire. *Quart. Journ. Geol. Soc.* vol. ix. p. 317.

——— (54). A Catalogue of British Fossils (2nd ed.). London, 1854.

Münster, G. Graf zu (43). Beiträge zur Petrefactenkunde. Bayreuth. Heft vi. Bayreuth.

Murchison, R. I. (34). Outline of the Geology of the neighbourhood of Cheltenham. Cheltenham, 1834.

——— *J. Buckman*, and *H. E. Strickland* (45). Outline of the Geology of the neighbourhood of Cheltenham. London, 1845.

Murray, G. (92). On a Fossil Alga belonging to the genus *Caulerpa*, from the Oolite. *Phycological Memoirs*, pt. i. p. 11, 1892.

Nathorst, A. G. (78–86). Om Floran i Skånes Kolförande Bildningar.

i. Floran vid Bjuf. Häft i. 1878.
ii. „ „ „ ii. 1879.
iii. „ „ „ iii. 1886.

Sver. Geol. Undersökn. Stockholm.

Nathorst, A. G. (78[1]). Beiträge zur fossilen flora Schwedens. Stuttgart, 1878.

——— (78[2]). Floran vid Höganäs och Helsingborg.

——— (80). Berättelse, abgifven till k. Vetenskaps Akademien, om en med understöd af allmänna medal utförd vetenskaplig resa till England. *Öfvers. k. Veten. Akad. Förhand.* No. 5, p. 23.

——— (02). Beiträge zur Kenntniss einiger Mesozoischen Cycadophyten. Ibid. vol. xxxvi. No. 4.

Newberry, J. S. (88). Rhætic Plants from Honduras. *Amer. Journ. Sci.* vol. xxxvi. p. 342.

Nicholson, H. A. and *R. Etheridge, jun.* (80). A monograph of the Silurian Fossils of the Girvan District in Ayrshire. Edinburgh.

Nicol, W. (34). Observations on the Structure of Recent and Fossil Coniferæ. *Edinburgh New Phil. Journ.* vol. xvi. p. 137.

Oldham, T. and *J. Morris* (63). Fossil Flora of the Gondwana System. Vol. i. pt. 1. *Mem. Geol. Surv. India*, ser. ii. Calcutta, 1880.

Parkinson, J. (11). Organic Remains of a Former World. London, 1811. (Vol. i. 1811 ; vol. ii. 1808 ; vol. iii. 1811.)

Penhallow, D. P. (00). Notes on the North American Species of *Dadoxylon*. *Trans. R. Soc. Canada* [2], vol. vi. p. 51.

Penzig, O. (94). Pflanzenteratologie. Vol. ii. Genua, 1894.

Phillips, J. (29). Illustrations of the Geology of Yorkshire. York, 1829.

——— (71). The Geology of Oxford and the Valley of the Thames. Oxford, 1871.

——— (75). Illustrations of the Geology of Yorkshire. Part i: The Yorkshire Coast (3rd ed.). Edited by R. Etheridge. London, 1875.

Plott, R. (1677). The Natural History of Oxfordshire. Oxford, 1677.

Pomel, A. (49). Matériaux pour servir à la Flore fossile des terrains jurassiques de la France. *Amt. Bericht. Versam. Ges. deutsch. Naturforsch. und Ärzte.* Aachen, 1849.

Porter, H. (63). On the occurrence of large quantities of Fossil Wood in the Oxford Clay near Peterborough. *Quart. Journ. Geol. Soc.* vol. xix. p. 317.

Potonié, H. (99). Lehrbuch der Pflanzenpalaeontologie. Berlin, 1899.

Prantl, K. (97). See *Engler & Prantl.*

Prestwich, J. (88). Geology. Vols. i.–ii. Oxford, 1888.

Quenstedt, F. A. (58). Der Jura. Tübingen, 1858.

Raciborski, M. (94). Flora Kopalna ogniotrwalych Glinek Krakowskich. *Pamiet. Mat. przy. Akad. Umiejetnósci.*

Renault, B. (81). Sur les *Sphenozamites*. *Compt. Rend.* Dec. 1881.

Richards, J. T. (84). Synopsis of British Fossil Cycadaceous Leaves.

——— (84[2]). On Scottish Fossil Cycadaceous Leaves contained in the Hugh Miller Collection. *Proc. R. Phys. Soc. Edinburgh*, 1884, p. 116.

Rothpletz, A. (91). Fossile Kalkalgen aus den Familien der Codiaceen und der Corallineen. *Zeit. deutsch. geol. Ges.* vol. xliii. p. 295, 1891.

Saporta, Le Marquis de (73–91). Paléontologie Française. Sér. ii. : Végétaux.—Plantes Jurassiques.

(73). Vol. i. : Algues, Equisétacées, Characées, Fougères.

(75). Vol. ii. : Cycadées.

(84). Vol. iii. : Conifères.

(91). Vol. v. : Types proangiospermiques, et supplément final.

——— (73[2]). Notice sur les plantes fossiles du niveau des lits à poissons de Cerin. Paris, 1873.

——— (94). Flore fossile du Portugal. *Dirèct. trav. géol. Portugal.* Lisbon, 1894.

Schenk, A. (66). Ueber die Flora der Schwarzen Schiefer von Raibl. *Würzb. nat. Zeitsch.* vol. vi. p. 10.

——— (67). Die fossile Flora der Grenzschichten der Keupers und Lias Frankens. Wiesbaden, 1867.

——— (83). Pflanzliche Versteinerungen. *Richthofen's* "China," vol. iv. Berlin, 1883.

——— (87). Fossile Pflanzen aus der Albourskette. *Bibl. bot. Uhlworm und Haenlein*, Heft vi. 1887.

Schimper, W. P. (69–74). Traité de paléontologie végétale.
Vol. i. 1869.
Vol. ii. 1870–72.
Vol. iii. 1874.

——— and *A. Mougeot* (44). Monographie des plantes fossiles du grès bigarré des Vosges. Leipzig, 1844.

Schütze, E. (01). Beiträge zur Kenntniss der triassischen Koniferen-Gattungen : *Pagiophyllum, Voltzia,* und *Widdringtonites. Jahresheft Ver. Vaterl. Naturkunde.* Stuttgart, 1901, p. 256.

Scott, D. H. (99). On the Primary Wood of certain Araucarioxylons. *Annals Bot.* vol. xiii. p. 615.

——— (00). Studies in Fossil Botany. London, 1900.

——— (02). On the Primary Structure of certain Palæozoic Stems with the *Dadoxylon* type of wood. *Trans. R. Soc. Edinburgh*, vol. xl. pt. 2, p. 331.

Sellards, E. H. (00). A new Genus of Ferns from the Permian of Kansas. *Kansas Univ. Quart.* vol. ix. p. 179.

Seward, A. C. (94). Catalogue of the Mesozoic Plants in the Department of Geology. British Museum (Nat. Hist.). The Wealden Flora, pt. i.

——— (94[2]). Algæ as Rock-building Organisms. *Science Progress,* vol. ii. p. 10.

——— (95). Catalogue of the Mesozoic Plants in the Department of Geology, British Museum. The Wealden Flora, pt. ii.

——— (96). Notes on the Geological History of the Monocotyledons. *Annals Bot.* vol. x. p. 205.

——— (97). The Glossopteris Flora. *Science Progress,* vol. i. (N.S.), p. 178.

——— (97[2]). On the Leaves of *Bennettites. Proc. Phil. Soc. Cambridge,* vol. ix. p. 273.

——— (98). Fossil Plants for Students of Botany and Geology. Vol. i. Cambridge, 1898.

——— (98[2]). On *Encephalartos Ghellinckii,* Lem. *Proc. Phil. Soc. Cambridge,* vol. ix. p. 340.

——— (99). On the Structure and Affinities of *Matonia pectinata,* R. Br., with notes on the Geological History of the Matonineæ. *Phil. Trans. R. Soc.* vol. cxci. p. 171.

——— (00). Catalogue of the Mesozoic Plants in the Department of Geology, British Museum. Jurassic Flora, vol. i.

——— (00[2]). Notes on some Jurassic Plants in the Manchester Museum. *Mem. and Proc. Manchester Lit. and Phil. Soc.* vol. xiv. pt. 3, p. 1.

——— (01). On the Structure and Origin of Jet. *Brit. Assoc. Rep.* p. 856, 1901.

——— (02). Article Palæobotany (Mesozoic). *Encyclopædia Britannica,* vol. xxxi. p. 421, 1902.

——— (03). Fossil Floras of Cape Colony. *Annals S. African Museum,* vol. iv.

——— (03[2]). Floras of the Past: their Composition and Distribution. Address to the Botanical Section of the British Association, Southport Meeting, 1903.

——— (03[3]). Occurrence of *Dictyozamites* in England. *Quart. Journ. Geol. Soc.* vol. lix. p. 217.

——— and *E. Dale* (01). On the Structure and Affinities of *Dipteris,* with notes on the Geological History of the Dipteridinæ. *Phil. Trans. R. Soc.* vol. cxciv. p. 487.

Seward, A. C. and *S. O. Ford* (03). The Anatomy of *Todea*, with notes on the Geological History and Affinities of the Osmundaceæ. *Trans. Linn. Soc.* vol. vi. pt. 5.

——— and *J. Gowan* (00). The Maidenhair Tree (*Ginkgo biloba*, L.). *Annals Bot.* vol. xiv. p. 109.

Sharp, S. (70). The Oolites of Northamptonshire. *Quart. Journ. Geol. Soc.* vol. xxvi. p. 354.

——— (73). The Oolites of Northamptonshire, pt. ii. *Quart. Journ. Geol. Soc.* vol. xxix, p. 225.

Siebold, F. de (70). Flora japonica. Vol. ii.

Sollas, I. B. J. (01). Fossils in the Oxford Museum. On the Structure and Affinities of the Rhætic plant *Naiadita*. *Quart. Journ. Geol. Soc.* vol. lvii. p. 307.

Solms-Laubach, Graf zu (91). Fossil Botany. English translation. Oxford, 1891.

——— (91). See *Capellini & Solms-Laubach.*

——— (91²). On the Fructification of *Bennettites Gibsonianus*, Carr. *Annals Bot.* vol. v. p. 419.

——— (99). Ueber das Genus *Pleuromeia*. *Bot. Zeit.* Heft xii. p. 227.

——— and *G. Steinmann* (99). Beiträge zur Geologie und Palaeontologie von Südamerika. *Neues Jahrb. Min. Beilageband*, xii. p. 593.

Sorby, H. C. (52). On the occurrence of a Non-Gymnospermous Exogenous Wood in the Lias near Bristol. *Trans. Micr. Soc. London*, vol. iii. p. 91.

Sternberg, C. Graf. von (20–38). Versuch einer geognostisch-botanischen Darstellung der Flora der Vorwelt. Leipzig.

Strahan, A. (98). The Geology of the Isle of Purbeck and Weymouth. *Mem. Geol. Surv.*

Stur, D. (85). Die obertriadische Flora der Lunzer-Schichten und des bituminosen Schiefer von Raibl. *Sitzb. k. Akad. Wiss. Wien.* vol. xci. Abth. 1.

Szajnocha, L. (88). Ueber fossile Pflanzenreste aus Cacheuta in den Argentinischen Republik. *Sitzb. k. Akad. Wiss. Wien.* vol. xcvii. Abth. 1, p. 219.

Tate, R. J. and *F. Blake* (76). The Yorkshire Lias. London, 1876.

Tenison-Woods, J. E. (83). On the Fossil Flora of the Coal Deposits of Australia. *Proc. Linn. Soc. N.S. Wales*, vol. viii. p. 37.

Thiselton-Dyer, W. T. (72). On some Coniferous Remains from the Lithographic Stone of Solenhofen. *Geol. Mag.* vol. ix. p. 150.

——— (78). Lecture on Plant Distribution as a field for Geographical Research. *Proc. Geogr. Soc.* vol. xxii. p. 412.

Unger, *F.* (45). Synopsis Plantarum fossilium. Lipseæ, 1845.

——— (47). Chloris Protogæa. Leipzig, 1847.

——— (50). Genera et species plantarum fossilium. Vindobonæ, 1850.

Velenovsky, *J.* (85). Die Gymnospermen der böhmischen Kreideformation. Prague, 1885.

Ward, *L. F.* (94). Fossil Cycadean Trunks of North America, with a Revision of the genus *Cycadeoidea*, Buckland. *Proc. Biol. Soc. Washington*, vol. ix. p. 75.

——— (95). The Potomac Formation. *Ann. Rep. Geol. Surv.* xv.

——— (96). Some Analogies in the Lower Cretaceous of Europe and America. *Ann. Rep. U.S. Geol. Surv.* 1896.

——— (00). Status of the Older Mesozoic Floras of the United States. *U.S. Geol. Surv.* 1900.

——— (00²). Description of a new genus and twenty new species of Fossil Cycadean Trunks from the Jurassic of Wyoming. *Proc. Washington Acad. Sci.* vol. i. p. 253.

Weiss, *E.* (86). Ueber eine Buntsandstein *Sigillaria* und deren nächste Verwandte. *Jahrb. k. preuss. Geol. Landes.* 1885.

Weld, *G.* (91). The Inferior Oolite of the Cotteswold Hills, with special reference to its Microscopical Structure. *Quart. Journ. Geol. Soc.* vol. xlvii. p. 550.

Wethered, *E. B.* (89). On the Microscopic Structure of the Jurassic Pisolite. *Geol. Mag.* vol. vi. p. 196.

——— (90). On the occurrence of *Girvanella* in Oolitic Rocks, and remarks on Oolitic Structure. *Quart. Journ. Geol. Soc.* vol. xlvi. p. 270.

——— (95). The Formation of Oolite. *Quart. Journ. Geol. Soc.* vol. li. p. 196.

Wickes, *W. H.* (00). A new Rhætic Section at Bristol. *Proc. Geol. Assoc.* vol. xvi. p. 421.

Wieland, *G. R.* (99). A Study of some American Fossil Cycads. *Amer. Journ. Sci.* vol. iv. p. 223.

——— (99²). Ibid. p. 389.

Williamson, *W. C.* (70). Contributions towards the History of *Zamia gigas*, L. & H. *Trans. Linn. Soc.* vol. xxvi. p. 663.

Wilson, *E.* (91). On a Section of the Rhætic Rocks at Pylle Hill (Totterdown), Bristol. *Quart. Journ. Geol. Soc.* vol. xlvii. p. 545.

Witham, *H.* (33). The Internal Structure of Fossil Vegetables found in the Carboniferous and Oolitic Deposits of Great Britain. Edinburgh, 1833.

Woodward, H. B. (76). The Geology of East Somerset and the Bristol Coalfields. *Mem. Geol. Surv.*

——— (93). The Jurassic Rocks of Britain. Vol. iii.: The Lias of England and Wales (Yorkshire excepted). *Mem. Geol. Surv.*

——— (94). The Jurassic Rocks of Britain. Vol. iv.: The Lower Oolitic Rocks of England (Yorkshire excepted). *Mem. Geol. Surv.*

——— (95). The Jurassic Rocks of Britain. Vol. v.: The Middle and Upper Oolitic Rocks of England (Yorkshire excepted). *Mem. Geol. Surv.* London, 1895.

Yokoyama, M. (89). Jurassic Plants from Kaga, Hida, and Echizen. *Journ. Coll. Sci. Japan*, vol. iii.

Young, G. (17). A History of Whitby and Streoneshalh Abbey. Whitby, 1817.

——— and *J. Bird* (22). A Geological Survey of the Yorkshire Coast. Whitby, 1822.

Zeiller, R. (82). Examen de la flore fossile des Couches de Charbon du Tong-King. *Ann. Mines*, vol. ii. [viii.], p. 299.

——— (83). Note sur la flore der bassin houiller de Tete. *Ann. Mines*, vol. iv. [viii.], p. 594.

——— (97). Les provinces botaniques de la fin des Temps primaires. *Rev. Gen. Sci.* 1897.

——— (00). Éléments de Paléobotanique. Paris, 1900.

——— (00^2). Sur les fossiles recueillis par M. Villiaume dans les roches charbonneuses des environs de Nossi-Bé. *Compt. rend.* June, 1900.

——— (02). Flore fossile des Gîtes de Charbon du Tonkin (Atlas). Paris, 1902.

——— (03). *Ibid.* Texte. Paris, 1903.

Zigno, A. de (56–85). Flora fossilis formationis Oolithicæ, vols. i. and ii. Padova, 1856.

Vol. i. (56).
Vol. ii. (73) Pt. 1, pp. 1–48 up to pl. xxxii.
„ (81) „ 2, pp. 49–80 up to pl. xxxiv.
„ (81) „ 3, pp. 81–120 up to pl. xl.
„ (85) „ 4, 5, pp. 121 to end.

Zittel, K. A. von (90). Handbuch der Palaeontologie. Abth. ii.: Schimper, W. P. and A. Schenk. Munich and Leipzig, 1890–1.

INDEX.

[Synonyms are printed in *italics*.]

A

Abies Webbiana, 60.
Acrostichites Phillipsii, 94.
Actinostrobus, 141.
Ægoceras planicostata, 52.
Africa, South, 2, 4.
Agassiz, A., 22.
Agathis, 131.
 australis, 121.
Algæ, 24, 26, 27, 76, 82.
Algites, 83.
 furcatus, 79, 83.
America, North, 1, 30, 42, 163.
America, South, 2, 153, 163.
Ammonites capricornus, 54.
 margaritatus, 54.
Andrae, K. T., 28.
Angiosperms, 4, 10, 11, 20, 21, 152–155.
Araucaria, 47, 49, 55, 130, 131, 136.
 Bidwilli, 137.
 brasiliensis, 49.
 Brodiei, 136.
 Cookii, 132, 136.
 excelsa, 48–51, 132, 136.
 Haeberleinii, 131.
 moreauana, 136.
 peregrina, 23, 48.
 sphærocarpa, 131.
Araucarieæ, 130–140.
Araucarioxylon, 55–59, 65.
 Lindleii, 56–61.
Araucarites, 20, 47, 75, 81, 123, 130–140, 170.
 Brodiei, 136–138.
 Bucklandi, 138.
 Cleminshawi, 139.
 Hudlestoni, 82, 133–135.
 ooliticus, 133–136, 138, 139.
 peregrinus, 48.
 sphæricus, 138.
 sphærocarpus, 131, 132, 135.
Aroides Stutterdi, 80, 155, 156.
Ashelworth, 10.
Athrotaxis, 141.
 cupressoides, 141.
Athrotaxites, 141.
Athrotaxopsis, 141.
Australian plants, 2, 34, 153, **163**.
Axminster, 23, 41.

B

Baiera, 101, 102.
 longifolia, 101.
 Phillipsi, 78, 99–102.
Baireuth, 9, 41.
Bajocian plants, 76.
Bamberg, 9.
Barber, C. A., 60.
Barnack, 76, 117, 148, 159.
Bartholin, C. T., 13.
Bath Museum, 41.
Bather, F. A., 156.
Bathonian plants, 76–81.
Bayer, E., 88.
Beche, H. T. de la, 23, 51, 52.
Bell, A. M., 119, 120, 145.
Bennettitales, 37, 43, 103–109.
Bennettites, 43, 45, 103–110.
 Gibsonianus, 43, 45, 103.
Bensonia, 122.
 ovata, 78, 80, 121.
Benstedtia, 44.
Berry, E. W., 30, 31.
Bessey, C. E., 153.
Binton, 20.
Biota orientalis, 146.
Blake, J. F., 54, 64, 82.
Blanckenhorn, M., 5.
Bornholm, 13.
Botanical provinces, 161, 170.
Bower, J. A., 64.

Brachyphyllum, 141, 147–150.
acutifolium, 142.
mamillare, 143, 146, 148.
nepos, 149.
peregrinum, 48.
solitarium, 25.
Braunston, 54, 158.
Bridgewater, 42.
Bristol, 11, 14, 23.
Bristol Museum, 41.
Brockeridge, 10.
Brodie, P. B., 10, 14, 15, 78.
Bromsgrove, 7.
Brongniart, A., 17, 83, 96, 108, 110, 117, 118.
Brown, A., 84, 85.
Bruton, 131, 159.
Buckland, W., 77, 83, 105, 106, 108–110, 139.
Bucklandia, 43–45, 109, 110.
squamosa, 80, 109.
Buckman, J., 10, 11, 16, 21, 78, 116, 119–122, 146, 147.
Bugbrook, 54.
Burcott Wood, 151.

C

Cædmon, 63.
Calamites, 78, 86.
Callitris, 141.
Cambrian rocks, 84.
Cambridge (Sedgwick Memorial) Museum, 87, 99, 102, 108, 129, 144.
Camptopteris Muensteriana, 18.
Capellini, G., 42.
Cardiff Museum, 41.
Carpolithes, 7, 19, 20, 75, 123–130, 139.
Bucklandi, 75, 124, 125, 127, 142.
conicus, 75, 78, 123–130, 139.
diospyriformis, 81, 123, 128–130, 170.
Lindleyanus, 81, 139.
Carruthers, W., 13, 15, 43, 45, 75, 81, 82, 106, 110, 124, 131, 133, 138, 149–151, 156.
Caulerpa, 85.
cactoides, 85.
Carruthersi, 82, 85, 86.
Caulerpites Bucklandianus, 142.
expansus, 142.
thuiæformis, 142.
Caulopteris tessellata, 5.
Ceratozamia mexicana, 111.
Chara, 26, 27.
liassica, 26.
Charmouth, 52, 54, 105, 158, 159.
Charterhouse, 26.
Chedworth, 85, 160.
Chili, 162.
China, 4, 122.
Chippenham, 149, 150, 160.
Chondrites bollensis, 26, 27.
Chordophyllites cicatricosus, 24, 26, 27.
Christian Malford, 81, 147–150, 160.
Cirencester Museum, 99, 100.
Cladophlebis denticulata, 169.
Clathraria Bucklandi, 109.
Clathropteris, 17–19, 162, 168.
meniscoides, 18.
platyphylla, 17–19, 25.
Cleeve Hill, 84, 159.
Coal, 75.
Collyweston, 76, 87–90, 159.
Commern, 5.
Coniferales, 47–62.
Coniferous wood, 52–62, 151.
Coniopteris, 93.
hymenophylloides, 93.
quinqueloba, 92.
Conites, 150, 151.
Bucklandi, 109.
depressus, 151.
primævus, 75, 150, 151.
Conwentz, H., 57.
Coral Rag, 151.
Corallian plants, 82, 160.
Coralline Oolite, 82, 130.
Corallineæ, 85.
Cordaites, 55.
Cracow, 90, 116.
Cretaceous plants, 30, 31, 112.
Crick, 53, 54.
Cryptomerites divaricatus, 146.
Ctenis, 35, 115, 116.
falcata, 116.
latifolia, 78, 79, 115, 116, 170.
orovallensis, 116.
Zeyschneri, 116.
Ctenopteris, 35–37.
cycadea, 25, 36, 37.
Sarrani, 37.
Ctenozamites, 35.
cycadea, 36.
Cunninghamia, 60.
Cupressineæ, 60, 140, 145.
Cupressinoxylon, 55, 60, 61.
Barberi, 60–62.
vectense, 60.
Cupressites liasinus, 50.

Cupressus, 140–143, 146.
latifolius, 10, 48.
Cyanophyceæ, 84.
Cycadella, 44.
Cycadeoidea, 42–47, 109, 110.
gracilis, 25, 45, 163.
pygmæa, 23, 25, 46, 47, 163.
squamosa, 80, 109, 110, 169.
Cycadeomyelon densecristatum, 110.
Cycadeospermum Bucklandi, 124.
conicum, 124.
Cycadeostrobus primævus, 150.
sphæricus, 81, 138.
Cycadinocarpus Bucklandi, 124.
conicus, 124.
Lindleyanus, 128.
Cycadites, 37–39, 78, 115, 11
Bucklandi, 109.
concentricus, 25, 38.
conferta, 37.
Delessei, 38.
pectinatus, 37.
plumula, 106.
rajmahalensis, 37.
rectangularis, 25, 37–39.
Saladini, 38.
Saportæ, 38.
tenuisectus, 38.
Cycadolepis, 21.
Cycadophyta, 37–47, 103–120, 169.
Cycadopteris, 29, 30.
Bergeri, 36.
Brauniana, 32, 33.
heterophylla, 32.
Cyclopteris latifolia, 79, 98.
longifolia, 101.
obtusa, 40.

D

Dacrydium, 60.
Dale, E., 89.
Damon, R., 75.
Darby Dale, 63.
Darwin, C., 154, 161.
De la Beche, H. T., 23, 51, 52.
Deane, H., 153.
Dichopteris, 28.
Dicotyledons, 10, 11, 20, 153.
Dictyophyllum, 17, 89, 90, 153–155, 168.
cracoviense, 90.
Dioonites, 38.
plumula, 107.
taxinus, 107.
Dipteridinæ, 17–19, 89, 90, 168.
Dipteris, 17, 19, 89.
conjugata, 90.
Distribution, Tables of, 164–167.
Drynaria, 88.
Dunker, W., 139.

E

Echinostipes pygmæus, 46.
Echinostrobus expansus, 143, 144.
Encephalartos, 110, 111.
Ghellinckii, 38.
pectinatus, 106.
taxinus, 106.
Equisetales, 12, 13, 27, 86, 168.
Equisetites, 12, 168.
Brodiei, 12, 26.
liassicum, 24.
Muensteri, 10, 12, 13, 26, 27, 162.
Equisetum Brodiei, 10.
pseudo-hoerense, 12.
Renaulti, 12, 13.
Sarrani, 13.
Ericaceæ, 10, 11, 20.
Eskdale, 63.
Etheridge, R., 76.
Ettingshausen, C. von, 27.
Eyeford, 77–79, 86, 93, 98, 100, 120, 122, 143, 146, 160.

F

Feistmantel, O., 108.
Filicales, 17–19, 27–35, 86–98.
Filicites, 78.
Bechei, 39.
Bucklandi, 40.
cycadea, 36.
meniscoides, 17.
scolopendroides, 106.
Fittonia, 43.
Fletton, 151.
Fliche, P., 5, 45, 134.
Fontaine, W. M., 141, 152.
Fontinalis, 15.
Ford, S. O., 131.
Forest Marble, 143, 144.
Fox-Strangways, C., 24, 54, 63, 64, 74.
Frič, A., & E. Bayer, 88.
Fucoides, 11.
elegans, 83.
furcatus, 83.
pennatula, 106, 108.
Fucoids, 26.
Fucus, 83.
Fungi, 82.

G

Gardner, J. S., 15, 122.
Geinitz, H. B., 163.
Geinitzia, 149.
Geological Survey Museum (London), *see* Jermyn Street.
Geyler, H. T., 122.
Ginkgo, 60, 98–100.
  biloba, 125, 126.
  digitata, 78, 80, 98-100.
Ginkgoales, 98–102, 169.
Ginkgodium Nathorsti, 100.
Girvanella, 76, 84.
  pisolitica, 84.
  problematica, 84.
Gleichenia lineata, 34.
Glenopteris, 28.
Glossopteris flora, 2.
Glossopteris longifolius, 79, 96.
  Phillipsi, 94.
Gnetales, 20.
Gnetum, 153.
Gondwana rocks, 2–4.
Göppert, H. R., 49.
Gorgonia Keuperi, 8.
Grantham, 54.
Great Oolite, 76–81, 159, 160.
Greenland, 13.
Gristhorpe Bay, 87.
Gymnosperm seeds, 123–130.
Gymnospermæ, 7, 19, 20, 98–152.

H

Halymenites ramulosus, 79, 83.
Hartz, N., 13.
Hippurites, 10, 11, 20, 21.
Hollick, A., 30.
Honduras, 119, 162, 163.
Horton, W. S., 78.
Hull, E., 78.
Huntcliff, 22.
Hutton, W., *see* Lindley & Hutton.
Hymenophyllites macrophyllus, 95, 96.
Hymenopteris macrophyllus, 79, 96.

I

India, Floras of, 2–4, 163.
Inferior Oolite System, 76, 100, 102, 139, 157.
Insect Limestone, 10, 14.
Italy, Floras of, 5, 28.

J

Jermyn Street Museum (London), 7, 41, 46, 82, 85, 88, 94, 117, 126, 129, 132, 143.
Jet, 54, 62–74.
Jet-rock, 22, 24.
Judd, J. W., 76, 117.
Juglandites castaneæfolius, 18.
Juniperus, 141.

K

Kaidacarpum, 133.
  ooliticum, 133.
Kellaways Rock, 130.
Kettering, 88.
Keuper flora, 5–7.
Keynsham, 11.
Kilsby, 53, 54, 158.
Kimeridge Clay, 82, 86, 151.
Kimeridge Coal, 75, 82.
Kimeridge plants, 82, 85, 160.
Kingsthorpe, 76, 87, 133, 134.
Kirby Moorside, 128.
Klukia exilis, 168.
Knowlton, F. H., 60, 65.
Krannera mirabilis, 111, 112.
Kurr, J. G., 23, 29, 49.
Kurtz, F., 153, 163.

L

Laccopteris, 86–89, 168.
  polypodioides, 76, 88.
  Woodwardi, 80, 87–89.
Langport, 42.
Leicester, 8.
Lesquereux, L., 30.
Lettenkohle flora, 6, 48.
Liassic System, 9–11, 22-72, 109, 110, 157, 163–168.
Libocedrus, 141.
Lignier, O., 104, 106, 110.
Lignite, 151.
Lilia, 122.
  lanceolata, 78, 80, 121, 122.
Lincolnshire Limestone, 107.
Lindley, J., & W. Hutton, 41, 47, 74, 75, 107, 124, 134, 139, 151.
Litchborough, 52, 54, 158.
Lithothamnion, 85.
Lomatopteris, 29, 30.
  jurensis, 26, 32.
Longdon, 7.

Lunéville, 5.
Lunz flora, 6.
Lycopodiales, 13–16.
Lycopodites, 13–16.
 falcatus, 16.
 lanceolatus, 10, 14–16.
Lyme Regis, 22, 23, 27, 33–42, 46, 47, 49–54, 158.
Lyons, H. G., 51.

M

McCoy, F., 93.
Macrotaeniopteris asplenoides, 26.
Macrozamia Fraseri, 126.
Madagascar, Jurassic plants of, 168.
Malton, 75, 82, 85, 123, 125, 126, 128, 130, 133, 134.
Malton Museum, 82, 125–128.
Manchester Museum, 75, 82, 116, 124, 127, 137, 144.
Mansel-Pleydell, J. C., 139.
Mantellia pygmæa, 47.
Marsiliaceæ, 93.
Matonia pectinata, 86, 88, 89.
 sarmentosa, 86.
Matonidium, 86, 168.
Matonineæ, 86–89, 168.
Megalosaurus, 77.
Microdictyon rutenicum, 87.
 Woodwardi, 87.
Möller, H., 13, 94.
Monocotyledons, 14, 20, 153, 156.
Moore, C., 26.
Moreauia acutifolia, 142.
 divaricata, 142.
 expansa, 142.
 latifolia, 48.
 podocarpoides, 95.
Morris, J., 29, 76, 163.
Moss, Supposed fossil, 15.
Mougeot, A., 4, 5.
Murray, G., 15, 82, 85.
Muschelkalk, plants of, 5.

N

Nageiopsis, 79, 152.
Naiadea, 14.
 acuminata, 14, 15.
 obtusa, 15, 78, 80, 119.
 ovata, 15, 78, 80, 120–122.
Naiadita acuminata, 10, 11.
 lanceolata, 13–15.
 petiolata, 10, 16.
Nathorst, A. G., 28, 29, 35, 37, 93, 104, 112.
Neuropteridium, 2, 4, 24.
Neuropteris limbata, 29.
New Zealand, 163.
Newberry, J. S., 162.
Nicol, W., 54, 55.
Nilssonia compta, 116.
Noeggerathia, 78, 98–100.
Noeggerathiopsis Hislopi, 112.
Nomenclature, 161, 162.
Northampton, 54.
Northampton Museum, 88, 133–135.
Nulliporites furcillatus, 24, 26, 27.

O

Odontopteris Bergeri, 36.
 Bucklandi, 40.
 cycadea, 36.
 jurensis, 29–33.
 latifolia, 115.
 Leckenbyi, 35.
 otopteris, 40.
Oleandridium vittatum, 91, 115.
Oolitic Series, 73–156, 158–160, 168–170.
Oroville (California), 116.
Otopteris acuminata, 10.
 Bucklandi, 40, 41.
 cuneata, 94.
 obtusa, 10, 23, 40.
Otozamites, 23, 39–42, 116, 117.
 acuminatus, 10, 25.
 Ameghinoi, 40.
 Beani, 118.
 Bechei, 40.
 brevifolius, var. *Bucklandi*, 40.
 Brongniarti, 40.
 gracilis, 25.
 graphicus, 41, 117.
 Hennoquei, 40.
 obtusus, 10, 23, 25, 39–42.
 obtusus, var. ooliticus, 116, 117.
Oxford Clay, 81, 138, 147–151.
Oxford Museum, 41, 75, 79–83, 87, 90–93, 96, 97, 105, 107, 108, 111–115, 119, 120, 137, 144, 145, 152.
Oxfordian plants, 81, 160.

P

Pachyphyllum liasinum, 25, 48.
 peregrinum, 24.

Pachypteris, 26, 28.
decurrens, 32.
lanceolata, 94.
thinnfeldia, 28.
Pagiophyllum, 47–51.
Foetterlei, 48.
peregrinum, 10, 23–25, 48–52.
Palæocyparis expansa, 143, 144.
flexuosa, 143.
Itieri, 143.
Palæozamia Bechei, 40.
Bucklandi, 40.
longifolia, 80, 110, 111, 114.
megaphylla, 80, 110–113.
pecten, 80, 107.
pectinata, 80, 106, 107.
taxina, 80, 106, 107.
Pandaneæ, 105, 133.
Pandanocarpum oolitícum, 133.
Paris (Yates Collection), 103.
Parkinson, J., 63.
Peach, C. W., 39.
Pecopteris approximata, 79, 92, 93.
diversa, 79, 152.
incisa, 79, 92, 93.
odontopteroides, 29, 163.
polypodioides, 76, 87.
Pendock, 7, 8.
Permian plants, 2, 118.
Permo-Carboniferous flora, 1-4.
Perucer Beds, 88.
Peterborough, 160.
Peuce Huttoniana, 24, 54.
Lindleii, 54, 56.
Lindleyana, 24, 54, 56.
Phillips, J., 79, 91, 107–111, 116, 139, 141, 145, 152.
Phlebopteris Woodwardi, 87.
Phyllites, 152–155.
Phyllocladus, 31, 60.
Phyllopteris Phillipsi, 94.
Phymatoderma liassicum, 26.
Pinites, 57, 78, 151.
dejectus, 151.
depressus, 82, 151.
elongatus, 51.
Lindleyanus, 56.
Pinus primæva, 75, 150.
Pitys primæva, 150.
Platycerium, 153.
Platylepis, 43.
Pleuromeia, 4, 5.
Pliny, 62.
Plott, R., 77.
Podocarpeæ, 60.
Podocarpus, 152.
Podozamites, 15, 119–123.
Podozamites lanceolatus, 14.
lanceolatus, var. geminus, 111.
lanceolatus, var. latifolius, 122.
Reinii, 122.
stonesfieldensis, 78, 80, 121, 170.
Poland, 13, 28.
Polypodiolites pectiniformis, 106, 108.
Ponton, 87, 89, 159.
Potomac plants, 112.
Proangiosperms, 155.
Protophyllocladus, 30, 31.
Protorhipis, 89, 90, 168.
Pterophyllum, 76.
Buckmanni, 87, 115.
comptum, 81.
giganteum, 112.
minus, 81.
Preslianum, 106, 107.
taxinum, 106, 107.
Ptilophyllum, 108.
acutifolium, 107.
pectinatum, 106.
taxina, 106.
Ptilozamites, 28, 29, 35.
Bergeri, 25, 36.
Leckenbyi, 25, 36.
Pylle Hill, 11.

R

Raciborski, M., 28, 90, 116.
Redland, 16.
Rhætic System, 9–23, 112, 119, 157, 161–163.
Rhodea macrophylla, 95.
Roade, 54, 158.
Rothpletz, A., 84.
Rothwell, 87, 88.
Rowington, 7, 8.
Rutland, 76.

S

Sagenopteris, 93, 94.
cuneata, 94.
Phillipsi, 94.
St. Honorine-la-Guillaume, 110.
Salicites longifolius, 78, 95–98.
Samaropsis Zignoana, 20.
Sandsfoot, 86, 160.
Saporta, Le Marquis de, 29, 36, 95, 103, 104, 106, 118, 144, 149.
Sarcostrobilus Paulini, 134.
Saxegothæa, 60.
Scania, 20, 28, 163.
Scarborough, 58, 130.
Scarborough Museum, 126.

Schenk, A., 12, 28, 41, 60, 122.
Schimper, W. P., 118, 139, 144.
Schimper & Mougeot, 4, 5.
Schizoneura, 2, 4.
gondwanensis, 4.
Schütze, E., 48.
Scott, D. H., 55.
Sedgwick Memorial Museum, *see* Cambridge.
Seeds, 19, 20, 123–130.
Sellards, E. H., 28.
Sevenhampton, 77–81, 86, 113–116, 122, 123, 130, 143, 147, 149, 160.
Sewardia, 118.
Sharp, S., 76.
Sherborne, 139, 159.
Sigillaria oculina, 5.
Siphoneæ, 82, 84.
Solenopora, 85.
jurassica, 85.
Sollas, I. B. J., 11, 13, 15, 16.
Sollas, W. J., 105.
Solms-Laubach, Graf zu, 42, 103, 104, 106, 163.
Sorby, H. C., 11.
Sphærococcites ramulosus, 83.
Sphenopteris, 75, 79, 92, 93.
cysteoides, 75, 79, 92.
elongata, 95.
longifolia, 101.
macrophylla, 95, 96.
plumosa, 79, 92, 93.
Sphenozamites, 117–120.
Belli, 78, 80, 119, 120, 170.
robustus, 119.
Rossii, 117.
Staithes, 27, 55, 58, 59, 158.
Stamford, 76, 87–89, 107, 108, 117.
Stangerites vittatus, 91.
Staverton, 54.
Steierdorf, 27, 32.
Stenopteris, 95, 97.
desmomera, 96, 97.
Sternberg, C. Graf von, 108, 109, 143, 144.
Stibbington, 117.
Stigmaria ficoides, 5.
Stigmarites Nicklesi, 5.
Stonesfield, 74–83, 86, 87, 90, 94, 95, 107–111, 113–115, 119–121, 129, 130, 137, 139, 140, 143, 144, 146–148, 152–154, 160.
Stonesfield Slate, 1, 78, 79, 81, 92, 93, 143, 151, 169, 170.
Stormberg Series, 95, 163.
Strahan, A., 75.
Strensham, 13.
Strickland, H. E., 77.
Strickland, Mrs., 99, 100, 102.
Strickland Collection (Cambridge), 99.
Stricklandia acuminata, 78, 80, 98–102.
Strobilites Bucklandi, 134, 138.
elongata, 23, 51, 52.
microphylla, 138.
Szajnocha, L., 163.

T

Tæniopteris, 91, 92.
angustata, 80, 87.
latifolia, 79, 115.
Phillipsii, 94.
scitaminea, 91.
scitamineæ-folia, 80, 91.
tenuinervis, 91.
vittata, 75, 80, 91, 92, 115.
Tate, R. J., & F. Blake, 54, 64.
Taxites podocarpioides, 81, 95, 96.
Taxodineæ, 60.
Tenison-Woods, J. E., 34.
Thallophyta, 26, 27, 82–86.
Thinnfeldia, 27–35, 95, 96.
decurrens, 32, 33.
falcata, 34.
incisa, 32.
lancifolia, 32.
obtusa, 32, 33.
odontopteroides, 29, 33, 163.
rhomboidalis, 26, 29, 31–33, 163.
speciosa, 78, 79, 81, 95–98, 169.
Thuites, 140–148.
acutifolia, 142.
articulatus, 81, 142–144.
cupressiformis, 78, 81, 142, 144.
divaricatus, 81, 142–144.
expansus, 78, 81, 137–148, 169.
pulchelliformis, 143.
Thuja, 140, 141, 143.
Thujopsis, 141.
Todia barbara, 168.
Todites Williamsoni, 168.
Tonkin plants, 13, 19, 37, 112, 163.
Totterdown, 11.
Towcester, 75, 151.
Trias plants, 1–5, 7, 8, 112, 157.

U

Ulospermum conicum, 124.
ornatum, 124.
Umbelliferæ, 10, 11, 20.
Unger, F., 18, 49.

V

Velenovsky, J., 112.
Voltzia, 8.
Vosges, Bunter plants of the, 4.

W

Wainlode Cliff, 10, 14, 15.
Wansford, 39, 107, 109, 117.
Ward, L. F., 1, 42–44, 112, 116.
Warkworth, 53.
Weekley, 87, 88.
Welwitschia, 20.
Wethered, E. B., 76, 84.
Weymouth, 82, 86, 151, 160.
Whitby, 52, 54, 55, 57–62, 65, 68, 158.
Whitby Museum, 27, 67, 68, 126.
Wickes, W. H., 11, 15, 16.
Wieland, G. R., 104, 106.
Williamson, W. C., 103–105.
Williamsonia, 103–109.
    Bucklandi, 105, 106.
    pecten, 25, 75, 80, 106–109.
Wilson, E., 9, 11.
Witham, H., 24, 54, 55, 65.
Withamia, 118.
Wolverton, 143, 144.
Wood, fossil coniferous, 52–62, 151.
Woodward, H. B., 11, 25, 74.
Wotton, 151.

X

Xanthosoma, 156.

Y

Yates Collection (Paris), 103.
Yatesia, 43–45.
    gracilis, 25, 45.
York Museum, 82, 126, 133, 135.
Young, G., 63.
Yuccites, 111, 112.
    megaphylla, 111.
    Schimperianus, 110, 112.

Z

Zamia, 44.
    gigas, 103.
    integrifolia, 118.
    muricata, 118.
    pectinata, 75, 106, 107.
    Skinneri, 118.
    taxina, 75, 106.
    Washingtoniana, 112.
Zamiostrobus primævus, 150.
    sphæricus, 138.
Zamites, 110–115.
    Bechei, 40.
    Bucklandi, 40.
    latifolius, 115.
    megaphyllus, 80, 110–115, 170.
    pectinata, 107.
    pygmæus, 46.
    taxina, 107.
Zeiller, R., 13, 29, 35, 112, 118, 141, 150, 163, 168.
Zigno, A. de, 29, 112, 117, 119.

PRINTED BY STEPHEN AUSTIN AND SONS, HERTFORD.

EXPLANATION OF PLATES.

WITH two exceptions (Pl. III. Fig. 5, and Pl. XII. Fig. 4), the figured specimens are preserved in the British Museum (Natural History), their registered numbers being quoted in square brackets. Except where otherwise stated, the figures are drawn natural size.

PLATE I.

FIG. 1. *Otozamites obtusus* (Lindley & Hutton). Page 41. [40,672.]

FIG. 2. *Otozamites obtusus* (L. & H.). P. 41. [1155.]

FIGS. 3, 3*a*. *Otozamites obtusus* (L. & H.). P. 42. [V. 22.]

FIG. 4. *Equisetites Muensteri*, Sternberg. P. 13. [V. 3358.]

FIG. 5. *Otozamites obtusus* (L. & H.). P. 42. [39,059.]

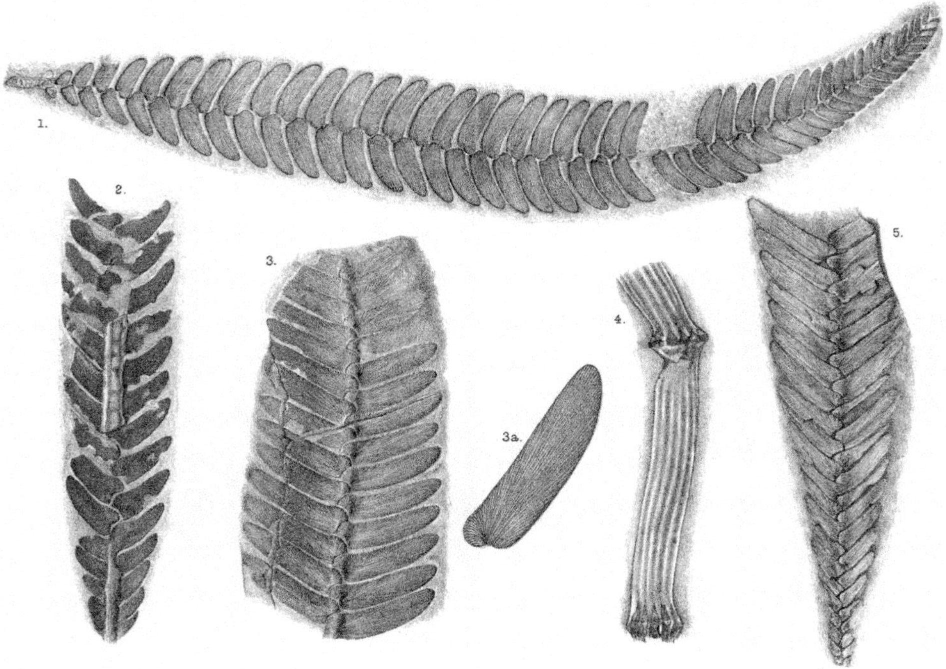

G.M.Woodward del et lith.

Figs 1–3, 5. Otozamites. 4. Equisetites.

West, Newman imp.

PLATE II.

FIG. 1. ? *Cycadites rectangularis*, Brauns. Page 39. [V. 9009.]

FIG. 2. *Lycopodites lanceolatus* (Brodie). P. 16. [V. 4015.]

FIG. 3. *Lycopodites lanceolatus* (Brodie). Part of a leaf enlarged to show the cells. P. 16. [V. 4015.]

1.

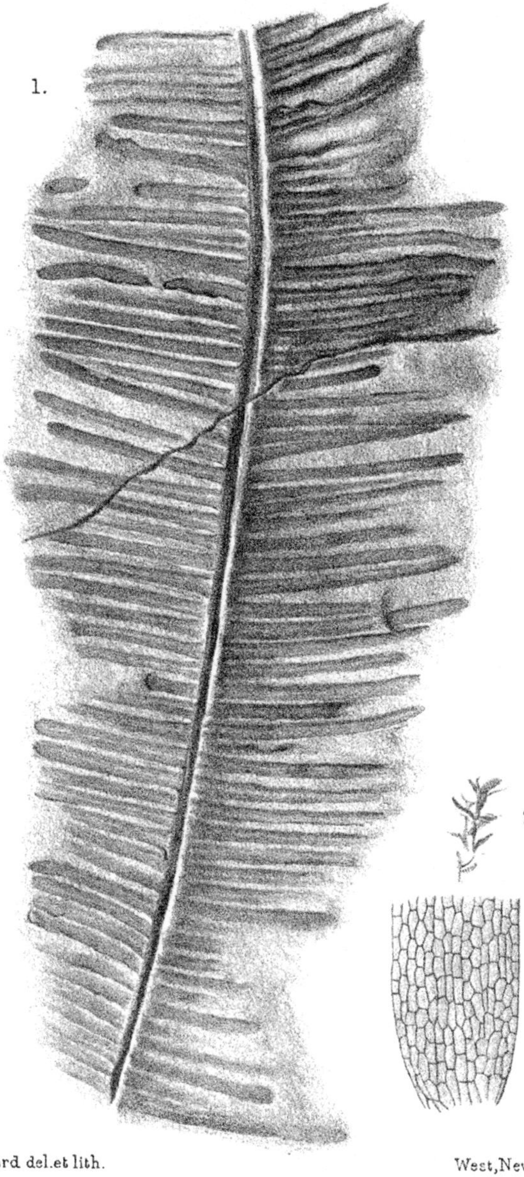

2.

3.

G.M. Woodward del. et lith. West, Newman imp.

Figs 1. Cycadites. 2, 3. Lycopodites.

PLATE III.

Figs. 1, 1*a*. *Ctenopteris cycadea*, Brongniart. Page 37. [40,674.]

Fig. 2. *Clathropteris platyphylla* (Göppert). P. 19. [V. 3399.]

Fig. 3. ? *Carpolithes conicus*, Lindley & Hutton. Internal cast. P. 128. [V. 9007.]

Fig. 4. *Podozamites stonesfieldensis*, sp. nov. P. 123. [V. 4662.]

Fig. 5. *Araucarites Brodiei*, Carruthers. P. 137. [Manchester Museum.]

G.M.Woodward del.et lith. West, Newman imp.

Figs 1, Ctenopteris. 2, Clathropteris.
5, Araucarites. Etc.

PLATE IV.

FIG. 1. *Thinnfeldia rhomboidalis*, Ettingshausen. Page 34. [52,751*a*.]

FIGS. 2, 2*a*. *Thinnfeldia rhomboidalis*, Ett. P. 34. [52,672.]

FIG. 3. *Thinnfeldia rhomboidalis*, Ett. P. 35. [52,751.]

FIGS. 4, 4*a*. *Cycadites rectangularis*, Brauns. P. 38. [52,665.]

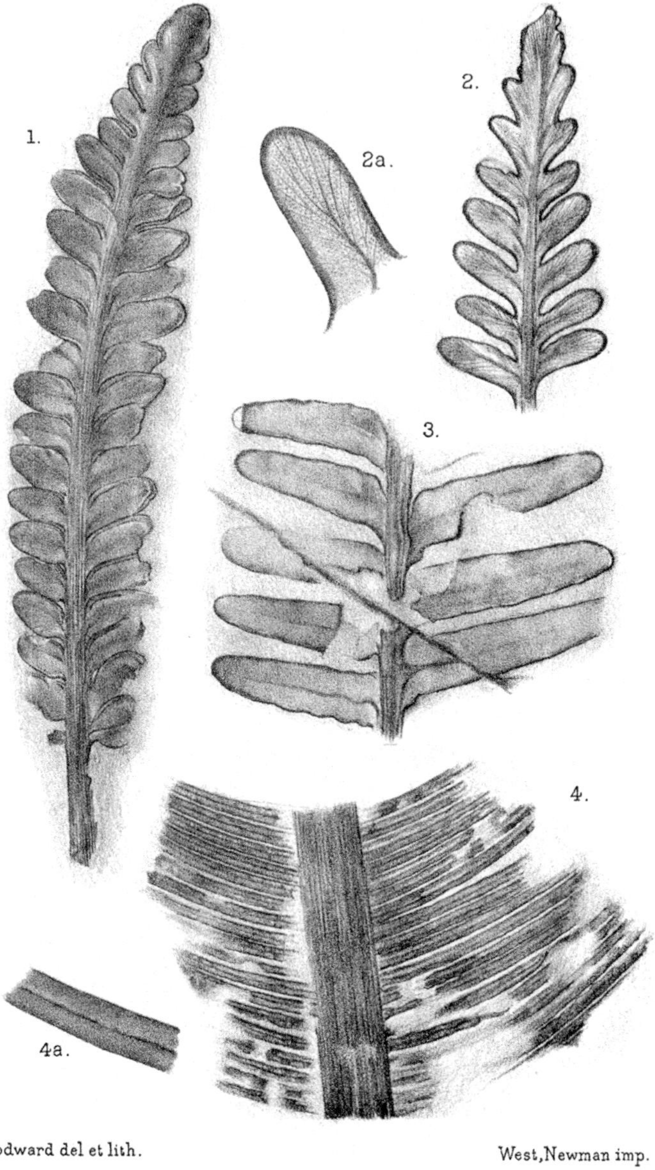

G.M.Woodward del et lith. West, Newman imp.

Figs 1–3, Thinnfeldia, 4, Cycadites.

PLATE V.

FIG. 1. *Pagiophyllum peregrinum* (Lindley & Hutton). Page 50. [38,349.]

FIG. 2. *Pagiophyllum peregrinum* (L. & H.). P. 50. [52,665*a*.]

FIG. 3. *Pagiophyllum peregrinum* (L. & H.). (× 5.) P. 50. [35,044.]

FIGS. 4, 4*a*. *Pagiophyllum peregrinum* (L. & H.). P. 50. [40,675.]

FIG. 5. *Pagiophyllum peregrinum* (L. & H.). P. 50. [V. 9608.]

FIG. 6. *Pagiophyllum peregrinum* (L. & H.). (× 5.) P. 51. [52,665.]

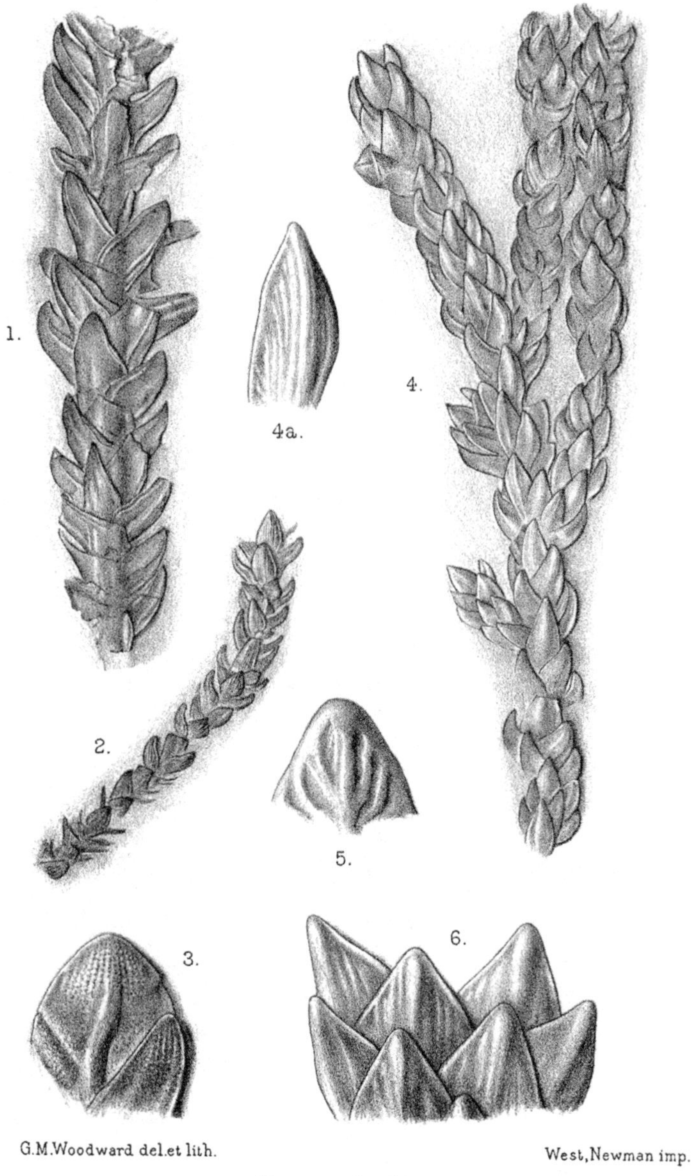

G.M. Woodward del. et lith. West, Newman imp.

Pagiophyllum.

PLATE VI.

Fig. 1. *Araucarioxylon Lindleii* (Witham). Page 56. [51,484.]

Fig. 2. *Araucarioxylon Lindleii* (With.). *c.* secretory tissue. P. 58. [51,724.]

Fig. 3. *Araucarioxylon Lindleii* (With.). P. 56. [51,484.]

Fig. 4. *Araucarioxylon Lindleii* (With.). P. 56. [51,484.]

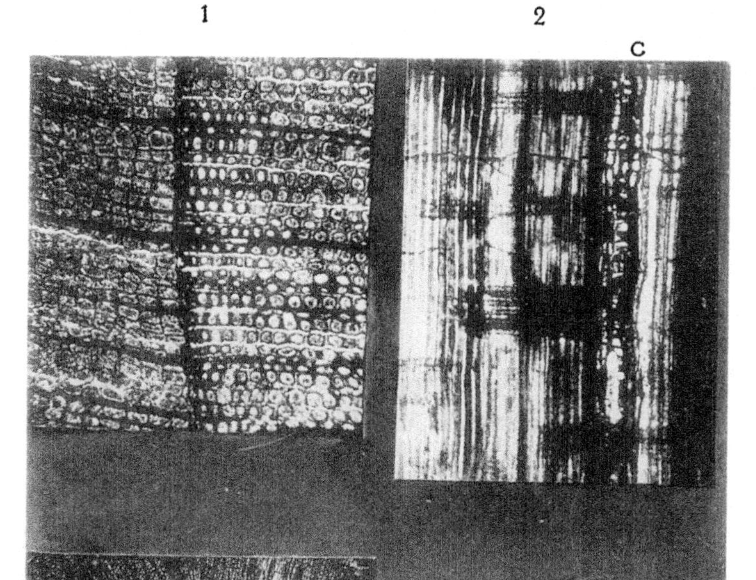

Araucarioxylon.

PLATE VII.

FIG. 1. *Cupressinoxylon Barberi*, sp. nov. Page 61. [51,498.]

FIG. 2. ? *Araucarioxylon Lindleii* (Witham). P. 59. [51,449.]

FIG. 3. *Araucarioxylon Lindleii* (With.). P. 56. [51,484.]

FIG. 4. *Cupressinoxylon Barberi*, sp. nov. (Slightly reduced.) P. 61. [51,496.]

FIG. 5. *Araucarioxylon Lindleii* (With.). P. 58. [51,488.]

FIG. 6. *Cupressinoxylon Barberi*, sp. nov. P. 61. [51,495.]

1, 4, 6, Cupressinoxylon. 2, 3, 5, Araucarioxylon

PLATE VIII.

Fig. 1. Jet and coniferous wood (*Araucarioxylon ?*). Page 68. [51,601.]

Fig. 2. Jet and coniferous wood. P. 68. [51,639.]

Fig. 3. Jet (pure) in tangential longitudinal section. P. 69. [51,595.]

Fig. 4. Jet and silicified wood (slightly reduced). P. 69. [51,493.]

Fig. 5. Jet (pure) in transverse section. P. 69. [51,623.]

Fig. 6. Jet and silicified wood (*Araucarioxylon ?*). P. 69. [51,448.]

Fig. 7. Jet (pure) in transverse section. P. 69. [51,623.]

1 2

6 7

Jet.

PLATE IX.

FIG. 1. *Thuites expansus*, Sternberg. Page 145. [40,513.]

FIG. 2. *Baiera Phillipsi*, Nathorst. P. 102. [V. 3433.]

FIG. 3. *Sagenopteris Phillipsi* (Brongniart). P. 94. [V. 4071.]

FIGS. 4, 4*a*. *Thuites expansus*, Sternb. Female flower. P. 147. [V. 3439.]

FIG. 5. *Brachyphyllum*, sp. *a*. P. 148. [11,130.]

FIG. 6. *Williamsonia pecten* (Phillips). P. 108. [V. 3361.]

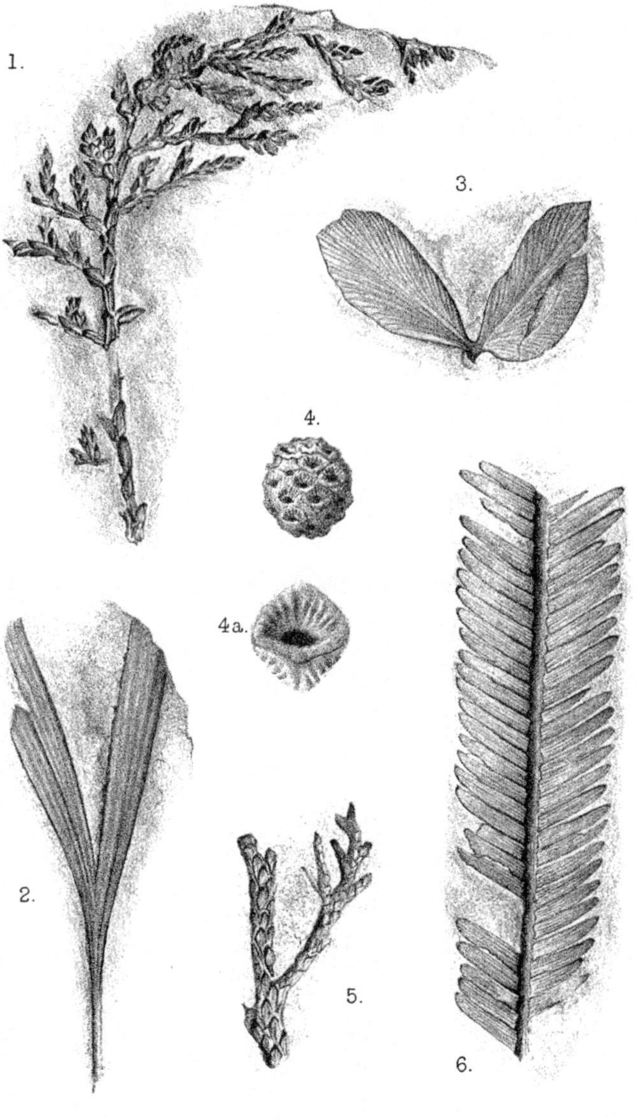

G.M. Woodward del. et lith. West, Newman imp.

Figs 1, 4, Thuites 3, Sagenopteris.
2, Baiera. 6, Williamsonia.

PLATE X.

FIG. 1. Cf. *Thinnfeldia speciosa*, Ettingshausen. Page 97. [V. 4074.]

FIG. 2. Cf. *Thinnfeldia speciosa*, Ett. P. 97. [V. 3422.]

FIG. 3. Cf. *Thinnfeldia speciosa*, Ett. P. 98. [V. 3425.]

FIG. 4. *Zamites megaphyllus* (Phillips). P. 112. [V. 86.]

FIG. 5. *Zamites megaphyllus* (Phill.). P. 113. [V. 4064.]

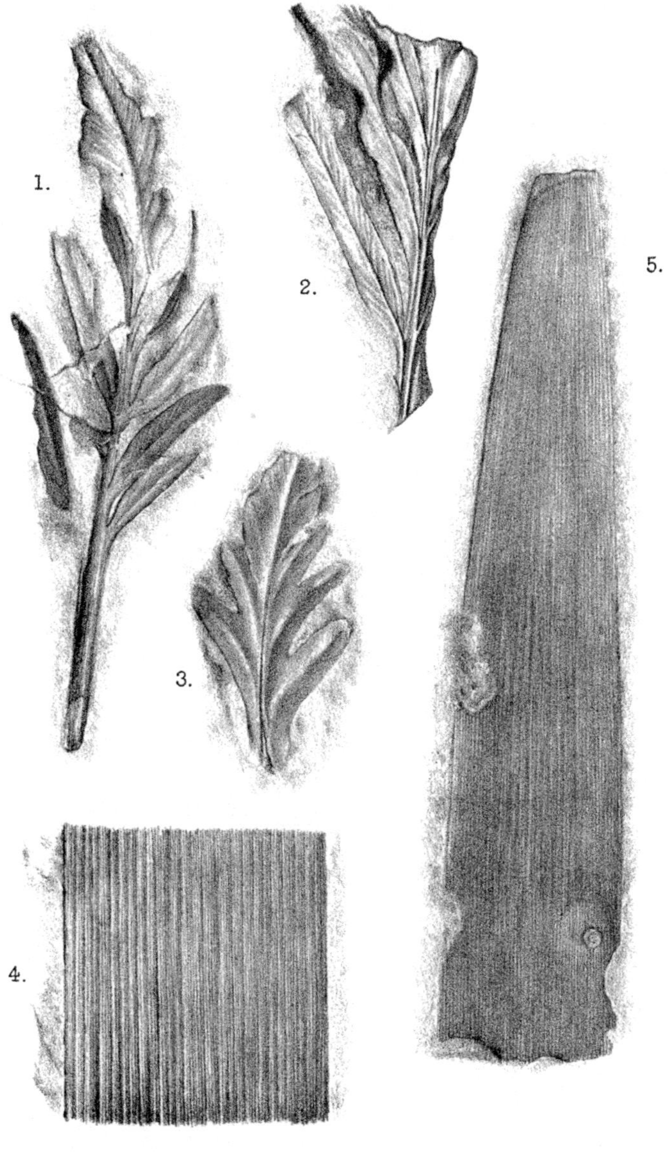

G.M.Woodward del.et lith. West,Newman imp.

Figs 1–3, Thinnfeldia. 4, 5, Zamites.

PLATE XI.

FIG. 1. *Podozamites stonesfieldensis*, sp. nov. Page 122. [V. 3420.]

FIG. 2. *Podozamites stonesfieldensis*, sp. nov. P. 122. [V. 3468.]

FIG. 3. *Ginkgo digitata* (Brongniart). P. 100. [V. 3429.]

FIG. 4. *Sphenozamites Belli*, sp. nov. P. 119. [V. 4069.]

FIGS. 5, 6. ? *Phyllites*, sp. P. 152. [41,385 and V. 85.]

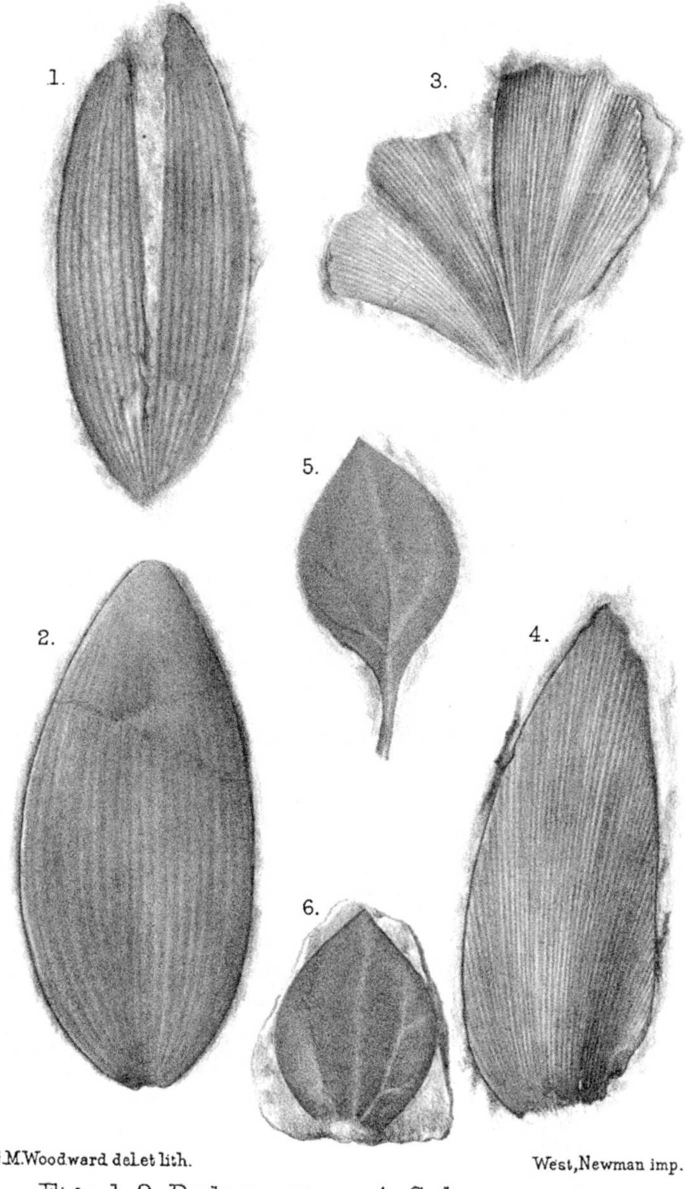

G.M. Woodward del. et lith.

West, Newman imp.

Figs 1, 2, Podozamites, 4, Sphenozamites.
3, Ginkgo. 5, 6, Phyllites.

PLATE XII.

Fig. 1. *Zamites megaphyllus* (Phillips). Page 113. [V. 3460.]

Fig. 2. *Araucarites Brodiei*, Carruthers. P. 137. [V. 3363.]

Fig. 3. *Zamites megaphyllus* (Phill.). P. 113. [V. 4652.]

Fig. 4. *Zamites megaphyllus* (Phill.). P. 113. [Oxford Museum.]

Fig. 5. *Zamites megaphyllus* (Phill.). P. 113. [V. 4067.]

Fig. 6. *Araucarites*, sp. P. 140. [V. 6591.]

Fig. 7. *Carpolithes diospyriformis*, Sternberg. P. 129. *s.* attachment-scar. [V. 2513*b*.]

Fig. 8. *Williamsonia pecten* (Phill.). P. 108. [52,868.]

Figs. 9, 9*a*. *Brachyphyllum*, sp. β (?). P. 149. [52,837.]

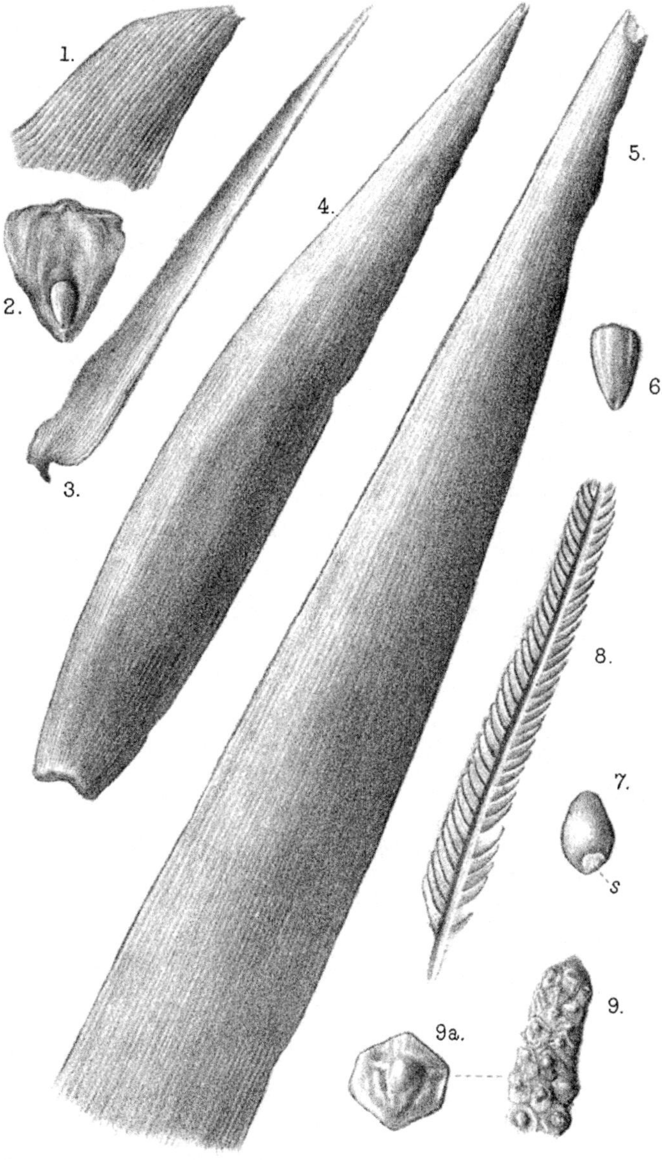

G. M. Woodward del. et lith. West, Newman imp.

Figs 1, 3, 4, 5, Zamites. 2, 6, Araucarites. Etc.

PLATE XIII.

FIG. 1. *Araucarites ooliticus* (Carruthers). Page 136. [52,840.]

FIG. 2. *Araucarites sphærocarpus*, Carr. Scale in longitudinal section. *a*. part of an adjacent scale ; *s*. seed. P. 132. [41,036.]

FIG. 3. *Araucarites sphærocarpus*, Carr. Distal ends of scales. *d*. transverse groove. P. 132. [41,036.]

FIG. 4. *Araucarites sphærocarpus*, Carr. Distal ends of scales. *d*. transverse groove. P. 132. [41,036.]

FIG. 5. *Carpolithes conicus*, Lindley & Hutton. *a*. median ridge. P. 126. [V. 9004.]

FIG. 6. *Carpolithes diospyriformis*, Sternberg. *s*. attachment - scar. P. 129. [V. 2513*a*.]

FIG. 7. *Carpolithes diospyriformis*, Sternb. *s*. attachment-scar. P. 129. [V. 2513.]

FIG. 8. *Araucarites sphærocarpus*, Carr Scale in surface - view. *d*. transverse groove ; *s*. position of seed ; *u*. umbo. P. 132. [41,036.]

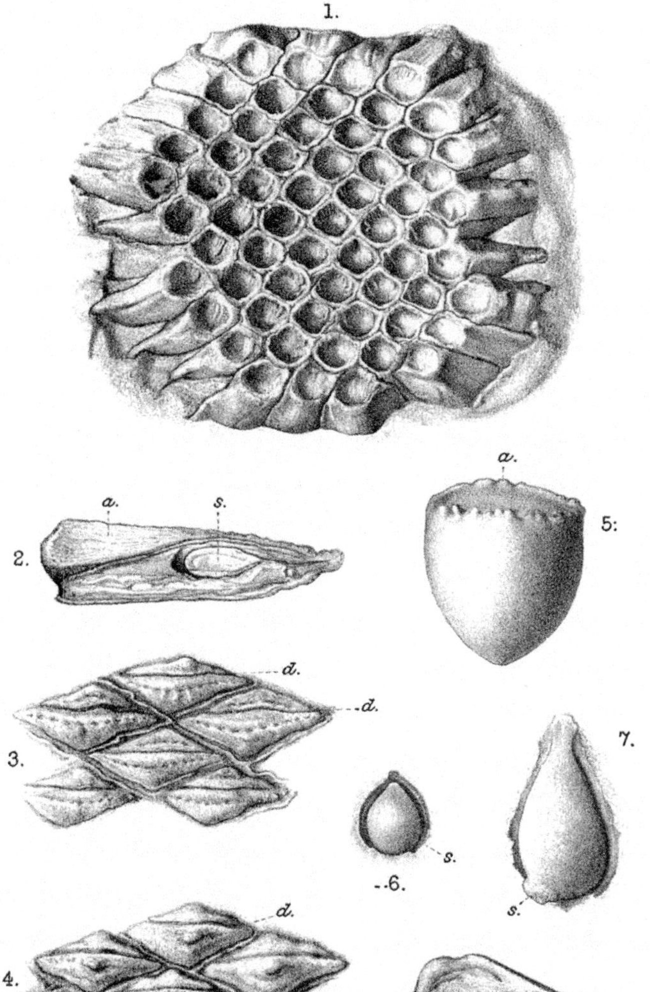

G.M.Woodward del.et lith. West, Newman imp.

Figs 1-4. 8, Araucarites, Figs 5-7, Carpolithes.

LIST OF THE CURRENT

NATURAL HISTORY PUBLICATIONS OF THE TRUSTEES OF THE BRITISH MUSEUM.

The following publications can be purchased through the Agency of *Messrs.* LONGMANS & CO., 39, *Paternoster Row*; *Mr.* QUARITCH, 15, *Piccadilly*; *Messrs.* KEGAN PAUL, TRENCH, TRÜBNER & CO., *Dryden House*, 43, *Gerrard Street, Soho, W.*; and *Messrs.* DULAU & CO., 37, *Soho Square*; or at the NATURAL HISTORY MUSEUM, *Cromwell Road, London, S.W.*

Catalogue of the Specimens and Drawings of Mammals, Birds, Reptiles, and Fishes of Nepal and Tibet. Presented by B. H. Hodgson, Esq., to the British Museum. 2nd edition. By John Edward Gray. Pp. xii., 90. [With an account of the Collection by Mr. Hodgson.] 1863, 12mo. 2*s.* 3*d.*

Catalogue of the Mammalia and Birds of New Guinea in the Collection of the British Museum. [With list of Species of New Guinea Birds, and those of the neighbouring Localities.] By John Edward Gray, Ph.D., F.R.S., and George Robert Gray, F.L.S., &c. Pp. 63. Woodcuts. 1859, 8vo. 1*s.* 6*d.*

Report on the Zoological Collections made in the Indo-Pacific Ocean during the voyage of H.M.S. "Alert," 1881–2. Pp. xxv., 684. 54 Plates. 1884, 8vo. 1*l.* 10*s.*

Summary of the Voyage...	By Dr. R. W. Coppinger.
Mammalia	„ O. Thomas.
Aves...	„ R. B. Sharpe.
Reptilia, Batrachia, Pisces	„ A. Günther.
Mollusca	„ E. A. Smith.
Echinodermata	„ F. J. Bell.
Crustacea	„ E. J. Miers.
Coleoptera	„ C. O. Waterhouse.
Lepidoptera	„ A. G. Butler.
Alcyonaria and Spongiida	„ S. O. Ridley.

A Monograph of Christmas Island (Indian Ocean): Physical Features and Geology by C. W. Andrews, B.A., B.Sc., F.G.S., with descriptions of the Fauna and Flora by numerous contributors. Pp. xv., 337: 22 plates (7 coloured), a map, and 27 illustrations in text. [With Index.] 1900, 8vo. 20*s.*

3500 Wt 12836 9/03 D & S 9 15699r A

Report on the Collections of Natural History made in the Antarctic Regions during the Voyage of the "Southern Cross." Pp. ix., 344. 53 Plates. 1902, Royal 8vo. 2*l.*

Mammalia	By	Capt. G. E. H. Barrett-Hamilton.
Notes on Antarctic Seals ...	,,	E. A. Wilson, M.B.
Extracts from the "Diary" of the late Nicolai Hanson.		——
Aves	,,	Dr. R. Bowdler Sharpe.
Pisces...	,,	G. A. Boulenger, F.R.S.
Tunicata	,,	Prof. W. A. Herdman, F.R.S.
Mollusca	,,	E. A. Smith.
Echinoderma	,,	Prof. F. J. Bell.
Insecta	,,	G. H. Carpenter and Hon. N. C. Rothschild.
Arachnida	,,	Dr. E. L. Trouessart.
Crustacea	,,	T. V. Hodgson.
Polychæta	,,	Dr. A. Willey.
Gephyrea	,,	A. E. Shipley.
Nematoda ... ... / Cestoda	,,	Dr. von Linstow.
Polyzoa ... ... / Porifera	,,	R. Kirkpatrick.
Anthozoa	,,	Dr. L. Roule and S. J. Hickson, F.R.S.
Actiniæ	,,	J. A. Clubb.
Hydrozoa	,,	E. T. Browne
Cryptogamia	,,	A. Gepp, V. H. Blackman, and Miss E. S. Barton.
Rock Specimens	,,	G. T. Prior.

Handbook of Instructions for Collectors, issued by the British Museum (Natural History). With Illustrations. Pp. 137. Index. 1902, 8vo. 2*s.* 6*d.*

The same, in ten parts, viz. :—Mammals; Birds; Reptiles, Batrachians, and Fishes; Insects; Diptera (Two-winged Flies); Mosquitoes (Culicidæ); Arachnida (Spiders, &c.), Myriopoda, and Peripatus; Soft-Bodied Invertebrates and Shells; Plants; Fossils and Minerals. 1900–1903, 8vo. 4*d.* each.

First Report on Economic Zoology. By Fred. V. Theobald, M.A., &c. [With an Introduction, containing a Classification of Animals from the point of view of Economic Zoology, by Prof. E. Ray Lankester, LL.D., F.R.S.] Pp. xxxiv., 192. 18 Woodcuts. 1903, Roy. 8vo. 6*s.*

Catalogue of the Books, Manuscripts, Maps, and Drawings in the British Museum (Natural History) :—

Vol. I. A—D. Compiled by B. B. Woodward, Assistant in charge of the General Library, with some clerical assistance. Pp. viii., 500. 1903, 4to. 20*s.*

MAMMALS.

Catalogue of the Bones of Mammalia in the Collection of the British Museum. By Edward Gerrard. Pp. iv., 296. 1862, 8vo. 5*s.*

Catalogue of Monkeys, Lemurs, and Fruit-eating Bats in the Collection of the British Museum. By Dr. J. E. Gray, F.R.S., &c. Pp. viii., 137. 21 Woodcuts. 1870, 8vo. 4*s.*

Catalogue of Carnivorous, Pachydermatous, and Edentate Mammalia in the British Museum. By John Edward Gray, F.R.S., &c. Pp. vii., 398. 47 Woodcuts. 1869, 8vo. 6*s*. 6*d*.

Catalogue of Seals and Whales in the British Museum. By John Edward Gray, F.R.S., &c. 2nd Edition. Pp. vii., 402. 101 Woodcuts. 1866, 8vo. 8*s*.

——— Supplement. By John Edward Gray, F.R.S., &c. Pp. vi., 103. 11 Woodcuts. 1871, 8vo. 2*s*. 6*d*.

List of the Specimens of Cetacea in the Zoological Department of the British Museum. By William Henry Flower, LL.D., F.R.S., &c. [With Systematic and Alphabetical Indexes.] Pp. iv., 36. 1885, 8vo. 1*s*. 6*d*.

Catalogue of Ruminant Mammalia (*Pecora*, Linnæus) in the British Museum. By John Edward Gray, F.R.S., &c Pp. viii., 102. 4 Plates. 1872, 8vo. 3*s*. 6*d*.

Catalogue of the Marsupialia and Monotremata in the Collection of the British Museum. By Oldfield Thomas. Pp. xiii., 401. 4 Coloured and 24 plain Plates. [With Systematic and Alphabetical Indexes.] 1888, 8vo. 1*l*. 8*s*.

BIRDS.

Catalogue of the Birds in the British Museum :—

Vol. VI. Catalogue of the Passeriformes, or Perching Birds, in the Collection of the British Museum. *Cichlomorphæ* : Part III., containing the first portion of the family Timeliidæ (Babbling Thrushes). By R. Bowdler Sharpe. Pp. xiii., 420. Woodcuts and 18 coloured Plates. [With Systematic and Alphabetical Indexes.] 1881, 8vo. 1*l*.

Vol. VII. Catalogue of the Passeriformes, or Perching Birds, in the Collection of the British Museum. *Cichlomorphæ* : Part IV., containing the concluding portion of the family Timeliidæ (Babbling Thrushes). By R. Bowdler Sharpe. Pp. xvi., 698. Woodcuts and 15 coloured Plates. [With Systematic and Alphabetical Indexes.] 1883, 8vo. 1*l*. 6*s*.

Vol. VIII. Catalogue of the Passeriformes, or Perching Birds, in the Collection of the British Museum. *Cichlomorphæ*: Part V., containing the families Paridæ and Laniidæ (Titmice and Shrikes) ; and *Certhiomorphæ* (Creepers and Nuthatches). By Hans Gadow, M.A., Ph.D. Pp. xiii., 386. Woodcuts and 9 coloured Plates. [With Systematic and Alphabetical Indexes.] 1883, 8vo. 17*s*.

Vol. IX. Catalogue of the Passeriformes, or Perching Birds, in the Collection of the British Museum. *Cinnyrimorphæ*, containing the families Nectariniidæ and Meliphagidæ (Sun Birds and Honey-eaters). By Hans Gadow, M.A., Ph.D. Pp. xii., 310. Woodcuts and 7 coloured Plates. [With Systematic and Alphabetical Indexes.] 1884, 8vo. 14*s*.

Catalogue of the Birds in the British Museum—*continued.*

Vol. X. Catalogue of the Passeriformes, or Perching Birds, in the Collection of the British Museum. *Fringilliformes*: Part I., containing the families Dicæidæ, Hirundinidæ, Ampelidæ, Mniotiltidæ, and Motacillidæ. By R. Bowdler Sharpe. Pp. xiii., 682. Woodcuts and 12 coloured Plates. [With Systematic and Alphabetical Indexes.] 1885, 8vo. 1*l.* 2*s.*

Vol. XI. Catalogue of the Passeriformes, or Perching Birds, in the Collection of the British Museum. *Fringilliformes*: Part II., containing the families Cœrebidæ, Tanagridæ, and Icteridæ. By Philip Lutley Sclater, M.A., F.R.S. Pp. xvii., 431. [With Systematic and Alphabetical Indexes.] Woodcuts and 18 coloured Plates. 1886, 8vo. 1*l.*

Vol. XII. Catalogue of the Passeriformes, or Perching Birds, in the Collection of the British Museum. *Fringilliformes*: Part III., containing the family Fringillidæ. By R. Bowdler Sharpe. Pp. xv., 871. Woodcuts and 16 coloured Plates. [With Systematic and Alphabetical Indexes.] 1888, 8vo. 1*l.* 8*s.*

Vol. XIII. Catalogue of the Passeriformes, or Perching Birds, in the Collection of the British Museum. *Sturniformes*, containing the families Artamidæ, Sturnidæ, Ploceidæ, and Alaudidæ. Also the families Atrichiidæ and Menuridæ. By R. Bowdler Sharpe. Pp. xvi., 701. Woodcuts and 15 coloured Plates. [With Systematic and Alphabetical Indexes.] 1890, 8vo. 1*l.* 8*s.*

Vol. XIV. Catalogue of the Passeriformes, or Perching Birds, in the Collection of the British Museum. *Oligomyodæ*, or the families Tyrannidæ, Oxyrhamphidæ, Pipridæ, Cotingidæ, Phytotomidæ, Philepittidæ, Pittidæ, Xenicidæ, and Eurylæmidæ. By Philip Lutley Sclater, M.A., F.R.S. Pp. xix., 494. Woodcuts and 26 coloured Plates. [With Systematic and Alphabetical Indexes.] 1888, 8vo. 1*l.* 4*s.*

Vol. XV. Catalogue of the Passeriformes, or Perching Birds, in the Collection of the British Museum. *Tracheophonæ*, or the families Dendrocolaptidæ, Formicariidæ, Conopophagidæ, and Pteroptochidæ. By Philip Lutley Sclater, M.A., F.R.S. Pp. xvii., 371. Woodcuts and 20 coloured Plates. [With Systematic and Alphabetical Indexes.] 1890, 8vo. 1*l.*

Vol. XVI. Catalogue of the Picariæ in the Collection of the British Museum. *Upupæ* and *Trochili*, by Osbert Salvin. *Coraciæ*, of the families Cypselidæ, Caprimulgidæ, Podargidæ, and Steatornithidæ, by Ernst Hartert. Pp. xvi., 703. Woodcuts and 14 coloured Plates. [With Systematic and Alphabetical Indexes.] 1892, 8vo. 1*l.* 16*s.*

Catalogue of the Birds in the British Museum—*continued.*

Vol. XVII. Catalogue of the Picariæ in the Collection of the British Museum. *Coraciæ* (contin.) and *Halcyones*, with the families Leptosomatidæ, Coraciidæ, Meropidæ, Alcedinidæ, Momotidæ, Totidæ and Coliidæ, by R. Bowdler Sharpe. *Bucerotes* and *Trogones* by W. R. Ogilvie Grant. Pp. xi., 522. Woodcuts and 17 coloured Plates. [With Systematic and Alphabetical Indexes.] 1892, 8vo. 1*l.* 10*s.*

Vol. XVIII. Catalogue of the Picariæ in the Collection of the British Museum. *Scansores*, containing the family Picidæ. By Edward Hargitt. Pp. xv., 597. Woodcuts and 15 coloured Plates. [With Systematic and Alphabetical Indexes.] 1890, 8vo. 1*l.* 6*s.*

Vol. XIX. Catalogue of the Picariæ in the Collection of the British Museum. *Scansores* and *Coccyges*: containing the families Rhamphastidæ, Galbulidæ, and Bucconidæ, by P. L. Sclater; and the families Indicatoridæ. Capitonidæ, Cuculidæ, and Musophagidæ, by G. E. Shelley. Pp. xii., 484: 13 coloured Plates. [With Systematic and Alphabetical Indexes.] 1891, 8vo. 1*l.* 5*s.*

Vol. XX. Catalogue of the Psittaci, or Parrots, in the Collection of the British Museum. By T. Salvadori. Pp. xvii., 658. Woodcuts and 18 coloured Plates. [With Systematic and Alphabetical Indexes.] 1891, 8vo. 1*l.* 10*s.*

Vol. XXI. Catalogue of the Columbæ, or Pigeons, in the Collection of the British Museum. By T. Salvadori. Pp. xvii., 676. 15 coloured Plates. [With Systematic and Alphabetical Indexes.] 1893, 8vo. 1*l.* 10*s.*

Vol. XXII. Catalogue of the Game Birds (*Pterocletes*, *Gallinæ*, *Opisthocomi*, *Hemipodii*) in the Collection of the British Museum. By W. R. Ogilvie Grant. Pp. xvi., 585. 8 coloured Plates. [With Systematic and Alphabetical Indexes.] 1893, 8vo. 1*l.* 6*s.*

Vol. XXIII. Catalogue of the Fulicariæ (Rallidæ and Heliornithidæ) and Alectorides (Aramidæ, Eurypygidæ, Mesitidæ, Rhinochetidæ, Gruidæ, Psophiidæ, and Otididæ) in the Collection of the British Museum. By R. Bowdler Sharpe. Pp. xiii., 353. 9 coloured Plates. [With Systematic and Alphabetical Indexes.] 1894, 8vo. 20*s.*

Vol. XXIV. Catalogue of the Limicolæ in the Collection of the British Museum. By R. Bowdler Sharpe. Pp. xii., 794. Woodcuts and 7 coloured Plates. [With Systematic and Alphabetical Indexes.] 1896, 8vo. 1*l.* 5*s.*

Catalogue of the Birds in the British Museum—*continued.*

Vol. XXV. Catalogue of the Gaviæ and Tubinares in the Collection of the British Museum. Gaviæ (Terns, Gulls, and Skuas), by Howard Saunders. Tubinares (Petrels and Albatrosses), by Osbert Salvin. Pp. xv., 475. Woodcuts and 8 coloured Plates. [With Systematic and Alphabetical Indexes.] 1896, 8vo. 1*l.* 1*s.*

Vol. XXVI. Catalogue of the Plataleæ, Herodiones, Steganopodes, Pygopodes, Alcæ, and Impennes in the Collection of the British Museum. Plataleæ (Ibises and Spoonbills) and Herodiones (Herons and Storks), by R. Bowdler Sharpe. Steganopodes (Cormorants, Gannets, Frigate-birds, Tropic-birds, and Pelicans), Pygopodes (Divers and Grebes), Alcæ (Auks), and Impennes (Penguins), by W. R. Ogilvie-Grant. Pp. xvii., 687. Woodcuts and 14 coloured Plates. [With Systematic and Alphabetical Indexes.] 1898, 8vo. 1*l.* 5*s.*

Vol. XXVII. Catalogue of the Chenomorphæ (Palamedeæ, Phoenicopteri, Anseres), Crypturi, and Ratitæ in the Collection of the British Museum. By T. Salvadori. Pp. xv., 636. 19 coloured Plates. [With Systematic and Alphabetical Indexes.] 1895, 8vo. 1*l.* 12*s.*

A Hand-list of the Genera and Species of Birds. [Nomenclator Avium tum Fossilium tum Viventium.] By R. Bowdler Sharpe, LL.D. :—

Vol. I. Pp. xxi., 303. [With Systematic Index.] 1899, 8vo. 10*s.*

Vol. II. Pp. xv., 312. [With Systematic Index, and an Alphabetical Index to Vols. I. and II.] 1900, 8vo. 10*s.*

Vol. III. Pp. xii., 367. [With Systematic and Alphabetical Indexes.] 1901, 8vo. 10*s.*

Vol. IV. Pp. xii., 391. [With Systematic and Alphabetical Indexes.] 1903, 8vo. 10*s.*

List of the Specimens of Birds in the Collection of the British Museum. By George Robert Gray :—

Part III., Section I. Ramphastidæ. Pp. 16. [With Index.] 1855, 12mo. 6*d.*

Part III., Section II. Psittacidæ. Pp. 110. [With Index.] 1859, 12mo. 2*s.*

Part III., Sections III. and IV. Capitonidæ and Picidæ. Pp. 137. [With Index.] 1868, 12mo. 1*s.* 6*d.*

Part IV. Columbæ. Pp. 73. [With Index.] 1856, 12mo. 1*s.* 9*d.*

Part V. Gallinæ. Pp. iv., 120. [With an Alphabetical Index.] 1867, 12mo. 1*s.* 6*d.*

Catalogue of the Birds of the Tropical Islands of the Pacific Ocean in the Collection of the British Museum. By George Robert Gray, F.L.S., &c. Pp. 72 [With an Alphabetical Index.] 1859, 8vo. 1*s.* 6*d.*

Catalogue of the Collection of Birds' Eggs in the British Museum (Natural History) :—

Vol. I. Ratitæ. Carinatæ (Tinamiformes—Lariformes). By Eugene W. Oates. Pp. xxiii., 252. 18 Coloured Plates. [With Systematic and Alphabetical Indexes.] 1901, 8vo. 30*s.*

Vol. II. Carinatæ (Charadriiformes—Strigiformes). By Eugene W. Oates. Pp. xx., 400. 15 Coloured Plates. [With Systematic and Alphabetical Indexes.] 1902, 8vo. 30*s.*

Vol. III. Carinatæ (Psittaciformes — Passeriformes). By Eugene W. Oates and Capt. Savile G. Reid. Pp. xxiii., 349. 10 Coloured Plates. [With Systematic and Alphabetical Indexes.] 1903, 8vo. 25*s.*

REPTILES.

Catalogue of the Tortoises, Crocodiles, and Amphisbænians in the Collection of the British Museum. By Dr. J. E. Gray, F.R.S., &c. Pp. viii., 80. [With an Alphabetical Index.] 1844, 12mo. 1*s.*

Catalogue of Shield Reptiles in the Collection of the British Museum. By John Edward Gray, F.R.S., &c.:—

Appendix. Pp. 28. 1872. 4to. 2*s.* 6*d.*

Part II. Emydosaurians, Rhynchocephalia, and Amphisbænians. Pp. vi., 41. 25 Woodcuts. 1872, 4to. 3*s.* 6*d.*

Hand-List of the Specimens of Shield Reptiles in the British Museum. By Dr. J. E. Gray, F.R.S., F.L.S., &c. Pp. iv., 124. [With an Alphabetical Index.] 1873, 8vo. 4*s.*

Catalogue of the Chelonians, Rhynchocephalians, and Crocodiles in the British Museum (Natural History). New Edition. By George Albert Boulenger. Pp. x., 311. 73 Woodcuts and 6 Plates. [With Systematic and Alphabetical Indexes.] 1889, 8vo. 15*s.*

Catalogue of the Lizards in the British Museum (Natural History). Second Edition. By George Albert Boulenger :—

Vol. I. Geckonidæ, Eublepharidæ, Uroplatidæ, Pygopodidæ, Agamidæ. Pp. xii., 436. 32 Plates. [With Systematic and Alphabetical Indexes.] 1885, 8vo. 20*s.*

Vol. II. Iguanidæ, Xenosauridæ, Zonuridæ, Anguidæ, Anniellidæ, Helodermatidæ, Varanidæ, Xantusiidæ, Teiidæ, Amphisbænidæ. Pp. xiii., 497. 24 Plates. [With Systematic and Alphabetical Indexes.] 1885, 8vo. 20*s.*

Vol. III. Lacertidæ, Gerrnosauridæ, Scincidæ, Anelytropidæ, Dibamidæ, Chamæleontidæ. Pp. xii., 575. 40 Plates. [With a Systematic Index and an Alphabetical Index to the three volumes.] 1887, 8vo. 1*l*. 6*s*.

Catalogue of the Snakes in the British Museum (Natural History). By George Albert Boulenger, F.R.S. :—

Vol. I., containing the families Typhlopidæ, Glauconiidæ, Boidæ, Ilysiidæ, Uropeltidæ, Xenopeltidæ, and Colubridæ aglyphæ, part. Pp. xiii., 448 : 26 Woodcuts and 28 Plates. [With Systematic and Alphabetical Indexes.] 1893, 8vo. 1*l*. 1*s*.

Vol. II., containing the conclusion of the Colubridæ aglyphæ. Pp. xi., 382 : 25 Woodcuts and 20 Plates. [With Systematic and Alphabetical Indexes.] 1894, 8vo. 17*s*. 6*d*.

Vol. III., containing the Colubridæ (Opisthoglyphæ and Proteroglyphæ), Amblycephalidæ, and Viperidæ. Pp. xiv., 727 : 37 Woodcuts and 25 Plates. [With Systematic Index, and Alphabetical Index to the 3 volumes.] 1896, 8vo. 1*l*. 6*s*.

Catalogue of Colubrine Snakes in the Collection of the British Museum. By Dr. Albert Günther. Pp. xvi., 281. [With Geographic, Systematic, and Alphabetical Indexes.] 1858, 12mo. 4*s*.

BATRACHIANS.

Catalogue of the Batrachia Salientia in the Collection of the British Museum. By Dr. Albert Günther. Pp. xvi., 160. 12 Plates. [With Systematic, Geographic, and Alphabetical Indexes.] 1858, 8vo. 6*s*.

FISHES.

Catalogue of the Fishes in the British Museum. Second edition. Vol. I. Catalogue of the Perciform Fishes in the British Museum. Vol. I. Containing the Centrarchidæ, Percidæ, and Serranidæ (part). By George Albert Boulenger, F.R.S. Pp. xix., 394. Woodcuts and 15 Plates. [With Systematic and Alphabetical Indexes.] 1895, 8vo. 15*s*.

Catalogue of Fish collected and described by Laurence Theodore Gronow, now in the British Museum. Pp. vii., 196. [With a Systematic Index.] 1854, 12mo. 3*s*. 6*d*.

Catalogue of Apodal Fish in the Collection of the British Museum. By Dr. Kaup. Pp. viii., 163. 11 Woodcuts and 19 Plates. 1856, 8vo. 10*s*.

Catalogue of Lophobranchiate Fish in the Collection of the British Museum. By J. J. Kaup, Ph.D., &c. Pp. iv., 80. 4 Plates. [With an Alphabetical Index.] 1856, 12mo. 2*s*.

MOLLUSCA.

Guide to the Systematic Distribution of Mollusca in the British Museum. Part I. By John Edward Gray, Ph.D., F.R.S., &c. Pp. xii., 230. 121 Woodcuts. 1857, 8vo. 5*s.*

List of the Shells of the Canaries in the Collection of the British Museum, collected by MM. Webb and Berthelot. Described and figured by Prof. Alcide D'Orbigny in the "Histoire Naturelle des Iles Canaries." Pp. 32. 1854, 12mo. 1*s.*

List of the Shells of Cuba in the Collection of the British Museum collected by M. Ramon de la Sagra. Described by Prof. Alcide d'Orbigny in the "Histoire de l'Ille de Cuba." Pp. 48. 1854, 12mo. 1*s.*

List of the Shells of South America in the Collection of the British Museum. Collected and described by M. Alcide D'Orbigny in the "Voyage dans l'Amérique Méridionale." Pp. 89. 1854, 12mo. 2*s.*

Catalogue of the Collection of Mazatlan Shells in the British Museum, collected by Frederick Reigen. Described by Philip P. Carpenter. Pp. xvi., 552. 1857, 12mo. 8*s.*

List of Mollusca and Shells in the Collection of the British Museum, collected and described by MM. Eydoux and Souleyet in the "Voyage autour du Monde, exécuté "pendant les années 1836 et 1837, sur la Corvette 'La "Bonite,'" and in the "Histoire naturelle des Mollusques "Ptéropodes." Par MM. P. C. A. L. Rang et Souleyet. Pp. iv., 27. 1855, 12mo. 8*d.*

Catalogue of the Phaneropneumona, or Terrestrial Operculated Mollusca, in the Collection of the British Museum. By Dr. L. Pfeiffer. Pp. 324. [With an Alphabetical Index.] 1852, 12mo. 5*s.*

Catalogue of Pulmonata, or Air Breathing Mollusca, in the Collection of the British Museum. Part I. By Dr. Louis Pfeiffer. Pp. iv., 192. Woodcuts. 1855, 12mo. 2*s.* 6*d.*

Catalogue of the Auriculidæ, Proserpinidæ, and Truncatellidæ in the Collection of the British Museum. By Dr. Louis Pfeiffer. Pp. iv., 150. Woodcuts. 1857, 12mo. 1*s.* 9*d.*

List of the Mollusca in the Collection of the British Museum. By John Edward Gray, Ph.D., F.R.S., &c.

Part II. Olividæ. Pp. 41. 1865, 12mo. 1*s.*

Catalogue of the Conchifera, or Bivalve Shells, in the Collection of the British Museum. By M. Deshayes :—

Part I. Veneridæ, Cyprinidæ, Glauconomidæ, and Petricoladæ. Pp. iv., 216. 1853, 12mo. 3*s.*

Part II. Petricoladæ (concluded); Corbiculadæ. Pp. 217–292. [With an Alphabetical Index to the two parts.] 1854, 12mo. 6*d.*

BRACHIOPODA.

Catalogue of Brachiopoda Ancylopoda or Lamp Shells in the Collection of the British Museum. [*Issued as* "Catalogue of the Mollusca, Part IV."] Pp. iv., 128. 25 Woodcuts. [With an Alphabetical Index.] 1853, 12mo. 3*s*.

POLYZOA.

Catalogue of Marine Polyzoa in the Collection of the British Museum. Part III. Cyclostomata. By George Busk, F.R.S. Pp. viii., 39. 38 Plates. [With a Systematic Index.] 1875, 8vo. 5*s*.

CRUSTACEA.

Catalogue of the Specimens of Amphipodous Crustacea in the Collection of the British Museum. By C. Spence Bate, F.R.S., &c. Pp. iv., 399. 58 Plates. [With an Alphabetical Index.] 1862, 8vo. 1*l*. 5*s*.

ARACHNIDA.

Descriptive Catalogue of the Spiders of Burma, based upon the Collection made by Eugene W. Oates and preserved in the British Museum. By T. Thorell. Pp. xxxvi., 406. [With Systematic List and Alphabetical Index.] 1895, 8vo. 10*s*. 6*d*.

INSECTS.

Coleopterous Insects.

Nomenclature of Coleopterous Insects in the Collection of the British Museum:—

Part VI. Passalidæ. By Frederick Smith. Pp. iv., 23. 1 Plate. [With Index.] 1852, 12mo. 8*d*.

Part VII. Longicornia, I. By Adam White. Pp. iv., 174. 4 Plates. 1853, 12mo. 2*s*. 6*d*.

Part VIII. Longicornia, II. By Adam White. Pp. 237. 6 Plates. 1855, 12mo. 3*s*. 6*d*.

Part IX. Cassididæ. By Charles H. Boheman, Professor of Natural History, Stockholm. Pp. 225. [With Index.] 1856, 12mo. 3*s*.

Illustrations of Typical Specimens of Coleoptera in the Collection of the British Museum. Part I. Lycidæ. By Charles Owen Waterhouse. Pp. x., 83. 18 Coloured Plates. [With Systematic and Alphabetical Indexes.] 1879, 8vo. 16*s*.

Catalogue of the Coleopterous Insects of Madeira in the Collection of the British Museum. By T. Vernon Wollaston, M.A., F.L.S. Pp. xvi., 234 : 1 Plate. [With a Topographical Catalogue and an Alphabetical Index.] 1857, 8vo. 3*s*.

Catalogue of the Coleopterous Insects of the Canaries in the Collection of the British Museum. By T. Vernon Wollaston, M.A., F.L.S. Pp. xiii., 648. [With Topographical and Alphabetical Indexes.] 1864, 8vo. 10*s*. 6*d*.

Catalogue of Halticidæ in the Collection of the British Museum. By the Rev. Hamlet Clark, M.A., F.L.S. Physapodes and Œdipodes. Part I. Pp. xii., 301. Frontispiece and 9 Plates. 1860, 8vo. 7*s*.

Catalogue of Hispidæ in the Collection of the British Museum. By Joseph S. Baly, M.E.S., &c. Part I. Pp. x., 172. 9 Plates. [With an Alphabetical Index.] 1858, 8vo. 6*s*.

Hymenopterous Insects.

Catalogue of Hymenopterous Insects in the Collection of the British Museum. By Frederick Smith. 12mo. :—

Part I. Andrenidæ and Apidæ. Pp. 197. 6 Plates. 1853, 2*s*. 6*d*.

Part II. Apidæ. Pp. 199–465. 6 Plates. [With an Alphabetical Index.] 1854. 6*s*.

Part III. Mutillidæ and Pompilidæ. Pp. 206. 6 Plates. 1855. 6*s*.

Part IV. Sphegidæ, Larridæ, and Crabronidæ. Pp. 207–497. 6 Plates. [With an Alphabetical Index.] 1856. 6*s*.

Part V. Vespidæ. Pp. 147. 6 Plates. [With an Alphabetical Index.] 1857. 6*s*.

Part VI. Formicidæ. Pp. 216. 14 Plates. [With an Alphabetical Index.] 1858. 6*s*.

Part VII. Dorylidæ and Thynnidæ. Pp. 76. 3 Plates. [With an Alphabetical Index.] 1859. 2*s*.

List of Hymenoptera, with descriptions and figures of the Typical Specimens in the British Museum. Vol. I., Tenthredinidæ and Siricidæ. By W. F. Kirby. Pp. xxviii., 450. 16 Coloured Plates. [With Systematic and Alphabetical Indexes.] 1882, 8vo. 1*l*. 18*s*.

Dipterous Insects.

List of the Specimens of Dipterous Insects in the Collection of the British Museum. By Francis Walker, F.L.S.

Part VII. Supplement III. Asilidæ. Pp. ii., 507–775. 1855, 12mo. 3*s*. 6*d*.

A Monograph of the Culicidæ, or Mosquitoes. Mainly compiled from the Collections received at the British Museum from various parts of the world in connection with the Investigation into the cause of Malaria conducted by the Colonial Office and the Royal Society. By Fred. V. Theobald, M.A., &c. :—

Vols. I., II. (pp. xviii., 424; viii., 391: 318 woodcuts), and Atlas of 37 coloured and 5 collotype plates (pp. viii., with explanations of plates). 1901, 8vo. 3*l*. 3*s*.

Vol. III. Pp. xvii., 359: 17 plates, 1 diagram, and 193 illustrations in text. 1903, 8vo. 1*l*. 1*s*.

A Monograph of the Tsetse-Flies (Genus Glossina, Westwood), based on the Collection in the British Museum. By Ernest Edward Austen. With a chapter on Mouth-parts by H. J. Hansen, Phil. Doc. Pp. ix., 319: 9 plates (7 coloured), 16 woodcuts, 1 map. 1903, Roy. 8vo. 15*s*.

Lepidopterous Insects.

Catalogue of the Lepidoptera Phalænæ in the British Museum. By Sir George F. Hampson, Bart. :—

Vol. I. Catalogue of the Syntomidæ in the Collection of the British Museum. Pp. xxi., 559: 285 Woodcuts. [With Systematic and Alphabetical Indexes.] 1898, 8vo. 15*s*.

——Atlas of 17 Coloured Plates, 8vo. 15*s*.

Vol. II. Catalogue of the Arctiadæ (Nolinæ, Lithosianæ) in the Collection of the British Museum. Pp. xx., 589. 411 Woodcuts. [With Systematic and Alphabetical Indexes.] 1900, 8vo. 18*s*.

——Atlas of 18 Coloured Plates (xviii.–xxxv.), 8vo. 15*s*.

Vol. III. Catalogue of the Arctiadæ (Arctianæ) and Agaristidæ in the Collection of the British Museum. Pp. xix., 690. 294 woodcuts. [With Systematic and Alphabetical Indexes.] 1901, 8vo. 15*s*.

——Atlas of 19 Coloured Plates (xxxvi–liv.), 8vo. 16*s*.

Illustrations of Typical Specimens of Lepidoptera Heterocera in the Collection of the British Museum :—

Part V. By Arthur Gardiner Butler. Pp. xii., 74. 78–100 Coloured Plates. [With a Systematic Index.] 1881, 4to. 2*l*. 10*s*.

Part VI. By Arthur Gardiner Butler. Pp. xv., 89. 101–120 Coloured Plates. [With a Systematic Index.] 1886, 4to. 2*l*. 4*s*.

Part VII. By Arthur Gardiner Butler. Pp. iv., 124. 121–138 Coloured Plates. [With a Systematic List.] 1889, 4to. 2*l*.

Part VIII. The Lepidoptera Heterocera of the Nilgiri District. By George Francis Hampson. Pp. iv., 144. 139–156 Coloured Plates. [With a Systematic List.] 1891, 4to. 2*l*.

Illustrations of Typical Specimens of Lepidoptera Heterocera in the Collection of the British Museum—*continued.*

Part IX. The Macrolepidoptera Heterocera of Ceylon. By George Francis Hampson. Pp. v., 182. 157-176 Coloured Plates. [With a General Systematic List of Species collected in, or recorded from, Ceylon.] 1893. 4to. 2*l*. 2*s*.

Catalogue of the Collection of Palæarctic Butterflies formed by the late John Henry Leech, and presented to the Trustees of the British Museum by his Mother, Mrs. Eliza Leech. By Richard South, F.E.S. Pp. vi., 228. 2 Coloured Plates. With a Portrait and Biographical Memoir of Mr. Leech. 1902, 4to. 1*l*.

Catalogue of Diurnal Lepidoptera described by Fabricius in the Collection of the British Museum. By Arthur Gardiner Butler, F.L.S., &c. Pp. iv., 303. 3 Plates. 1869, 8vo. 7*s*. 6*d*.

Specimen of a Catalogue of Lycænidæ in the British Museum. By W. C. Hewitson. Pp. 15. 8 Coloured Plates. 1862, 4to. 1*l*. 1*s*.

List of Lepidopterous Insects in the Collection of the British Museum. Part I. Papilionidæ. By G. R. Gray, F.L.S. Pp. 106. [With an Alphabetical Index.] 1856, 12mo. 2*s*.

List of the Specimens of Lepidopterous Insects in the Collection of the British Museum. By Francis Walker. 12mo. :—

Part XXII. Geometrites. Pp. 499-755. 1861, 3*s*. 6*d*.
Part XXIII. ——— Pp. 756-1020. 1861, 3*s*. 6*d*.
Part XXIV. ——— Pp. 1021-1280. 1862, 3*s*. 6*d*.
Part XXV. ——— Pp. 1281-1477. 1862, 3*s*.
Part XXVI. ——— Pp. 1478-1796. [With an Alphabetical Index to Parts XX.-XXVI.] 1862, 4*s*. 6*d*.
Part XXVII. Crambites and Tortricites. Pp. 1-286. 1863, 4*s*.
Part XXVIII. Tortricites and Tineites. Pp. 287-561. 1863, 4*s*.
Part XXIX. Tineites. Pp. 562-835. 1864, 4*s*.
Part XXX. ——— Pp. 836-1096. [With an Alphabetical Index to Parts XXVII.-XXX.] 1864, 4*s*.
Part XXXI. Supplement. Pp. 1-321. 1864, 5*s*.
Part XXXII. ——— Part 2. Pp. 322-706. 1865, 5*s*.
Part XXXIII. ——— Part 3. Pp. 707-1120. 1865, 6*s*.
Part XXXIV. ——— Part 4. Pp. 1121-1533. 1865, 5*s*. 6*d*.
Part XXXV. ——— Part 5. Pp. 1534-2040. [With an Alphabetical Index to Parts XXXI.-XXXV.] 1866, 7*s*.

Neuropterous Insects.

Catalogue of the Specimens of Neuropterous Insects in the Collection of the British Museum. By Francis Walker. 12mo. :—

Part II. Sialidæ—Nemopterides. Pp. ii., 193–476. 1853, 3*s.* 6*d.*

Part III. Termitidæ—Ephemeridæ. Pp. ii., 477–585. 1853, 1*s.* 6*d.*

Catalogue of the Specimens of Neuropterous Insects in the Collection of the British Museum. By Dr. H. Hagen. Part I. Termitina. Pp. 34. 1858, 12mo. 6*d.*

Orthopterous Insects.

Catalogue of Orthopterous Insects in the Collection of the British Museum. Part I. Phasmidæ. By John Obadiah Westwood, F.L.S., &c. Pp. 195. 48 Plates. [With an Alphabetical Index.] 1859, 4to. 3*l.*

Catalogue of the Specimens of Blattariæ in the Collection of the British Museum. By Francis Walker, F.L.S., &c. Pp. 239. [With an Alphabetical Index.] 1868, 8vo. 5*s.* 6*d.*

Catalogue of the Specimens of Dermaptera Saltatoria [Part I.] and Supplement to the Blattariæ in the Collection of the British Museum. Gryllidæ. Blattariæ. Locustidæ. By Francis Walker, F.L.S., &c. Pp. 224. [With an Alphabetical Index.] 1869, 8vo. 5*s.*

Catalogue of the Specimens of Dermaptera Saltatoria in the Collection of the British Museum. By Francis Walker, F.L.S., &c.—

Part II. Locustidæ (continued). Pp. 225–423. [With an Alphabetical Index.] 1869, 8vo. 4*s.* 6*d.*

Part III. Locustidæ (continued).—Acrididæ. Pp. 425–604. [With an Alphabetical Index.] 1870, 8vo. 4*s.*

Part IV. Acrididæ (continued). Pp. 605–809. [With an Alphabetical Index.] 1870, 8vo. 6*s.*

Part V. Tettigidæ.—Supplement to the Catalogue of Blattariæ.—Supplement to the Catalogue of Dermaptera Saltatoria (with remarks on the Geographical Distribution of Dermaptera). Pp. 811–850; 43; 116. [With Alphabetical Indexes.] 1870, 8vo. 6*s.*

Hemipterous Insects.

Catalogue of the Specimens of Heteropterous Hemiptera in the Collection of the British Museum. By Francis Walker, F.L.S., &c. 8vo. :—

Part I. Scutata. Pp. 240. 1867. 5*s.*

Part II. Scutata (continued). Pp. 241–417. 1867. 4*s.*

Part III. Pp. 418–599. [With an Alphabetical Index to Parts I., II., III., and a Summary of Geographical Distribution of the Species mentioned.] 1868. 4*s.* 6*d.*

Part IV. Pp. 211. [With Alphabetical Index.] 1871. 6*s.*

Part V. Pp. 202. [With Alphabetical Index.] 1872. 5*s.*

Part VI. Pp. 210. [With Alphabetical Index.] 1873. 5*s.*

Part VII. Pp. 213. [With Alphabetical Index.] 1873. 6*s.*

Part VIII. Pp. 220. [With Alphabetical Index.] 1873. [6*s.* 6*d.*

VERMES.

Catalogue of the Species of Entozoa, or Intestinal Worms, contained in the Collection of the British Museum. By Dr. Baird. Pp. iv., 132. 2 Plates. [With an Index of the Animals in which the Entozoa mentioned in the Catalogue are found, and an Index of Genera and Species.] 1853, 12mo. 2*s.*

ANTHOZOA.

Catalogue of Sea-pens or Pennatulariidæ in the Collection of the British Museum. By J. E. Gray, F.R.S., &c. Pp. iv., 40. 2 Woodcuts. 1870, 8vo. 1*s.* 6*d.*

Catalogue of Lithophytes or Stony Corals in the Collection of the British Museum. By J. E. Gray, F.R.S., &c. Pp. iv., 51. 14 Woodcuts. 1870, 8vo. 3*s.*

Catalogue of the Madreporarian Corals in the British Museum (Natural History) :—

Vol. I. The Genus Madrepora. By George Brook. Pp. xi., 212. 35 Collotype Plates. [With Systematic and Alphabetical Indexes, and Explanation of the Plates.] 1893, 4to. 1*l.* 4*s.*

Vol. II. The Genus Turbinaria ; the Genus Astræopora. By Henry M. Bernard, M.A. Cantab., F.L.S., F.Z.S. Pp. iv., 106. 30 Collotype and 3 Lithographic Plates. [With Index of Generic and Specific Names, and Explanation of the Plates.] 1896, 4to. 18*s.*

Vol. III. The Genus Montipora; the Genus Anacropora. By Henry M. Bernard, M.A. Pp. vii., 192. 30 Collotype and 4 Lithographic Plates. [With Systematic Index, Index of Generic and Specific Names, and Explanation of the Plates.] 1897, 4to. 1*l*. 4*s*.

Vol. IV. The Family Poritidæ. I.—The Genus Goniopora. By Henry M. Bernard, M.A. Pp. viii., 206. 12 Collotype and 4 Lithographic Plates. [With Index of Generic and Specific Names, and Explanation of the Plates.] 1903, 4to. 1*l*.

BRITISH ANIMALS.

Catalogue of British Birds in the Collection of the British Museum. By George Robert Gray, F.L.S., F.Z.S., &c. Pp. xii., 248. [With a List of Species.] 1863, 8vo. 3*s*. 6*d*.

Catalogue of British Hymenoptera in the Collection of the British Museum. Second edition. Part. I. Andrenidæ and Apidæ. By Frederick Smith, M.E.S. New Issue. Pp. xi., 236. 11 Plates. [With Systematic and Alphabetical Indexes.] 1891, 8vo. 6*s*.

Catalogue of British Fossorial Hymenoptera, Formicidæ, and Vespidæ in the Collection of the British Museum. By Frederick Smith, V.P.E.S. Pp. 236. 6 Plates. [With an Alphabetical Index.] 1858, 12mo. 6*s*.

A Catalogue of the British Non-parasitical Worms in the Collection of the British Museum. By George Johnston, M.D., Edin., F.R.C.L., Ed., LL.D., Marischal Coll., Aberdeen, &c. Pp. 365. Woodcuts and 24 Plates. [With an Alphabetical Index.] 1865, 8vo. 7*s*.

Catalogue of the British Echinoderms in the British Museum (Natural History). By F. Jeffrey Bell, M.A. Pp. xvii., 202. Woodcuts and 16 Plates (2 Coloured). [With Table of Contents, Tables of Distribution, Alphabetical Index, Description of the Plates, &c.] 1892, 8vo. 12*s*. 6*d*.

List of the Specimens of British Animals in the Collection of the British Museum; with Synonyma and References to figures. 12mo.:—

Part V. Lepidoptera. By J. F. Stephens. 2nd Edition. Revised by H. T. Stainton and E. Shepherd. Pp. iv., 224. 1856. 1*s*. 9*d*.

Part VI. Hymenoptera. By F. Smith. Pp. 134. 1851. 2*s*.

Part VII. Mollusca, Acephala and Brachiopoda. By Dr. J. E. Gray. Pp. iv., 167. 1851. 3*s*. 6*d*.

Part VIII. Fish. By Adam White. Pp. xxiii., 164. (With Index and List of Donors.) 1851. 3*s.* 6*d.*

Part IX. Eggs of British Birds. By George Robert Gray. Pp. 143. 1852. 2*s.* 6*d.*

Part XI. Anoplura, or Parasitic Insects. By H. Denny. Pp. iv., 51. 1852. 1*s.*

Part XII. Lepidoptera (continued). By James F. Stephens. Pp. iv., 54. 1852. 9*d.*

Part XIII. Nomenclature of Hymenoptera. By Frederick Smith. Pp. iv., 74. 1853, 1*s.* 4*d.*

Part XIV. Nomenclature of Neuroptera. By Adam White. Pp. iv., 16. 1853, 6*d.*

Part XV. Nomenclature of Diptera, I. By Adam White. Pp. iv., 42. 1853. 1*s.*

Part XVI. Lepidoptera (completed). By H. T. Stainton. Pp. 199. [With an Index.] 1854. 3*s.*

Part XVII. Nomenclature of Anoplura, Euplexoptera, and Orthoptera. By Adam White. Pp. iv., 17. 1855. 6*d.*

PLANTS.

Illustrations of the Botany of Captain Cook's Voyage Round the World in H.M.S. "Endeavour" in 1768–71. By the Right Hon. Sir Joseph Banks, Bart., K.B., P.R.S., and Dr. Daniel Solander, F.R.S. With Determinations by James Britten, F.L.S., Senior Assistant, Department of Botany, British Museum :—

Part I.—Australian Plants. 101 Copper-plates [after paintings by F. P. Nodder], with 31 pages of descriptive text. 1900, fol. 25*s.*

Part II.—Australian Plants, Pt. II. 142 Copper-plates (pls. 101–243) [after paintings by F. P. Nodder, James Miller, J. F. Miller, and John Cleveley], with 41 pages of descriptive text (pp. 35–75). 1901, fol. 35*s.*

Catalogue of the African Plants collected by Dr. Friedrich Welwitsch in 1853–61 :—

Vol. I. Dicotyledons. By William Philip Hiern, M.A. F.L.S., &c. :—

Part I. [Ranunculaceæ to Rhizophoraceæ.] Pp. xxvi., 336. [With Portrait of Dr. Welwitsch. Introduction, Bibliography, and Index of Genera.] 1896, 8vo. 7*s.* 6*d.*

Part II. Combretaceæ to Rubiaceæ. Pp. 337–510. [With Index of Genera.] 1898, 8vo. 4*s.*

Part III. Dipsaceæ to Scrophulariaceæ. Pp. 511–784. [With Index of Genera.] 1898, 8vo. 5*s.*

Part IV. Lentibulariaceæ to Ceratophylleæ. Pp. 785–1035. [With Index.] 1900, 8vo. 5*s.*

Vol II. Monocotyledons, Gymnosperms, and Cryptogams :—

Part I. Monocotyledons and Gymnosperms. By Alfred Barton Rendle, M.A., D.Sc. F.L.S., Assistant, Department of Botany. Pp. 260. [With Index of Genera.] 1899, 8vo. 6*s.*

Part II. Cryptogamia. Pp. 261–566. [With Table of Errata, and General Index to the whole work.] 1901, 8vo. 6*s.*

Vascular Cryptogams ...	By	William Carruthers, F.R.S.
Mosses	„	Antony Gepp, M.A., F.L.S.
Hepatics	„	F. Stephani.
Marine Algæ	„	Ethel S. Barton.
Freshwater Algæ	„	W. West, F.L.S., and G. S. West, B.A.
Diatomaceæ	„	Thomas Comber, F.L.S.
Lichenes	„	E. A. Wainio.
Fungi	„	Annie Lorrain Smith.
Mycetozoa	„	Arthur Lister, F.R.S.

A Monograph of Lichens found in Britain : being a Descriptive Catalogue of the Species in the Herbarium of the British Museum. By the Rev. James M. Crombie, M.A., F.L.S., F.G.S., &c. Part I. Pp. viii., 519 : 74 Woodcuts. [With Glossary, Synopsis, Tabular Conspectus, and Index.] 1894, 8vo. 16*s.*

A Monograph of the Mycetozoa : being a Descriptive Catalogue of the Species in the Herbarium of the British Museum. By Arthur Lister, F.L.S. Pp. 224. 78 Plates and 51 Woodcuts. [With Synopsis of Genera and List of Species, and Index.] 1894, 8vo. 15*s.*

List of British Diatomaceæ in the Collection of the British Museum. By the Rev. W. Smith, F.L.S., &c. Pp. iv., 55. 1859, 12mo. 1*s.*

FOSSILS.

Catalogue of the Fossil Mammalia in the British Museum (Natural History). By Richard Lydekker, B.A., F.G.S. :—

Part I. Containing the Orders Primates, Chiroptera, Insectivora, Carnivora, and Rodentia. Pp. xxx., 268. 33 Woodcuts. [With Systematic and Alphabetical Indexes.] 1885, 8vo. 5*s*.

Part II. Containing the Order Ungulata, Suborder Artiodactyla. Pp. xxii., 324. 39 Woodcuts. [With Systematic and Alphabetical Indexes.] 1885, 8vo. 6*s*.

Part III. Containing the Order Ungulata, Suborders Perissodactyla, Toxodontia, Condylarthra, and Amblypoda. Pp. xvi., 186. 30 Woodcuts. [With Systematic Index, and Alphabetical Index of Genera and Species, including Synonyms.] 1886, 8vo. 4*s*.

Part IV. Containing the Order Ungulata, Suborder Proboscidea. Pp. xxiv., 235. 32 Woodcuts. [With Systematic Index, and Alphabetical Index of Genera and Species, including Synonyms.] 1886, 8vo. 5*s*.

Part V. Containing the Group Tillodontia, the Orders Sirenia, Cetacea, Edentata, Marsupialia, Monotremata, and Supplement. Pp. xxxv., 345. 55 Woodcuts. [With Systematic Index, and Alphabetical Index of Genera and Species, including Synonyms.] 1887, 8vo. 6*s*.

Catalogue of the Fossil Birds in the British Museum (Natural History). By Richard Lydekker, B.A. Pp. xxvii., 368. 75 Woodcuts. [With Systematic Index, and Alphabetical Index of Genera and Species, including Synonyms.] 1891, 8vo. 10*s*. 6*d*.

Catalogue of the Fossil Reptilia and Amphibia in the British Museum (Natural History). By Richard Lydekker, B.A., F.G.S. :—

Part I. Containing the Orders Ornithosauria, Crocodilia, Dinosauria, Squamata, Rhynchocephalia, and Proterosauria. Pp. xxviii., 309. 69 Woodcuts. [With Systematic Index, and Alphabetical Index of Genera and Species, including Synonyms.] 1888, 8vo. 7*s*. 6*d*.

Part II. Containing the Orders Ichthyopterygia and Sauropterygia. Pp. xxi., 307. 85 Woodcuts. [With Systematic Index, and Alphabetical Index of Genera and Species, including Synonyms.] 1889, 8vo. 7*s*. 6*d*.

Part III. Containing the Order Chelonia. Pp. xviii., 239. 53 Woodcuts. [With Systematic Index, and Alphabetical Index of Genera and Species, including Synonyms.] 1889, 8vo. 7*s.* 6*d.*

Part IV. Containing the Orders Anomodontia, Ecaudata, Caudata, and Labyrinthodontia; and Supplement. Pp. xxiii., 295. 66 Woodcuts. [With Systematic Index, Alphabetical Index of Genera and Species, including Synonyms, and Alphabetical Index of Genera and Species to the entire work.] 1890, 8vo. 7*s.* 6*d.*

Catalogue of the Fossil Fishes in the British Museum (Natural History). By Arthur Smith Woodward, LL.D., F.R.S., F.G.S., &c. :—

Part I. Containing the Elasmobranchii. Pp. xlvii., 474. 13 Woodcuts and 17 Plates. [With Alphabetical Index, and Systematic Index of Genera and Species.] 1889, 8vo. 21*s.*

Part II. Containing the Elasmobranchii (Acanthodii), Holocephali, Ichthyodorulites, Ostracodermi, Dipnoi, and Teleostomi (Crossopterygii and Chondrostean Actinopterygii). Pp. xliv., 567. 58 Woodcuts and 16 Plates. [With Alphabetical Index, and Systematic Index of Genera and Species.] 1891, 8vo. 21*s.*

Part III. Containing the Actinopterygian Teleostomi of the Orders Chondrostei (concluded), Protospondyli, Aetheospondyli, and Isospondyli (in part). Pp. xlii., 544. 45 Woodcuts and 18 Plates. [With Alphabetical Index, and Systematic Index of Genera and Species.] 1895, 8vo. 21*s.*

Part IV. Containing the Actinopterygian Teleostomi of the Suborders Isospondyli (in part), Ostariophysi, Apodes, Percesoces, Hemibranchii, Acanthopterygii, and Anacanthini. Pp. xxxix., 636. 22 Woodcuts and 19 Plates. [With Alphabetical Index, and Systematic Index of Genera and Species.] 1901, 8vo. 21*s.*

Systematic List of the Edwards Collection of British Oligocene and Eocene Mollusca in the British Museum (Natural History), with references to the type-specimens from similar horizons contained in other collections belonging to the Geological Department of the Museum. By Richard Bullen Newton, F.G.S. Pp. xxviii., 365. [With table of Families and Genera, Bibliography, Correlation-table, Appendix, and Alphabetical Index.] 1891, 8vo. 6*s.*

Catalogue of Tertiary Mollusca in the Department of Geology, British Museum (Natural History). Part I. The Australasian Tertiary Mollusca. By George F. Harris, F.G.S., &c. Pp. xxvi., 407. 8 Plates. [With Table of Families, Genera, and SubGenera, and Index.] 1897, 8vo. 10*s.*

Catalogue of the Fossil Cephalopoda in the British Museum (Natural History) :—

Part I. Containing part of the Suborder Nautiloidea, consisting of the families Orthoceratidæ, Endoceratidæ, Actinoceratidæ, Gomphoceratidæ, Ascoceratidæ, Poterioceratidæ, Cyrtoceratidæ, and Supplement. By Arthur H. Foord, F.G.S. Pp. xxxi., 344. 51 Woodcuts. [With Systematic Index, and Alphabetical Index of Genera and Species, including Synonyms.] 1888, 8vo. 10*s*. 6*d*.

Part II. Containing the remainder of the Suborder Nautiloidea, consisting of the families Lituitidæ, Trochoceratidæ, Nautilidæ, and Supplement. By Arthur H. Foord, F.G.S. Pp. xxviii., 407. 86 Woodcuts. [With Systematic Index, and Alphabetical Index of Genera and Species, including Synonyms.] 1891, 8vo. 15*s*.

Part III. Containing the Bactritidæ, and part of the Suborder Ammonoidea. By Arthur H. Foord, Ph.D., F.G.S., and George Charles Crick, A.R.S.M., F.G.S. Pp. xxxiii., 303. 146 Woodcuts. [With Systematic Index of Genera and Species, and Alphabetical Index.] 1897, 8vo. 12*s*. 6*d*.

List of the Types and Figured Specimens of Fossil Cephalopoda in the British Museum (Natural History). By G. C. Crick, F.G.S. Pp. 103. [With Index.] 1898, 8vo. 2*s*. 6*d*.

A Catalogue of British Fossil Crustacea, with their Synonyms and the Range in Time of each Genus and Order. By Henry Woodward, F.R.S. Pp. xii., 155. [With an Alphabetical Index.] 1877, 8vo. 5*s*.

Catalogue of the Fossil Bryozoa in the Department of Geology, British Museum (Natural History):—

The Jurassic Bryozoa. By J. W. Gregory, D.Sc., F.G.S., F.Z.S. Pp. [viii.,] 239 : 22 Woodcuts and 11 Plates. [With List of Species and Distribution, Bibliography, Index, and Explanation of Plates.] 1896, 8vo. 10*s*.

The Cretaceous Bryozoa. Vol. I. By J. W. Gregory, D.Sc., F.G.S., F.Z.S. Pp. xiv., 457 : 64 Woodcuts and 17 Plates. [With Index and Explanation of Plates.] 1899, 8vo. 16*s*.

Catalogue of the Blastoidea in the Geological Department of the British Museum (Natural History), with an account of the morphology and systematic position of the group, and a revision of the genera and species. By Robert Etheridge, jun., of the Department of Geology, British Museum (Natural History), and P. Herbert Carpenter, D.Sc., F.R.S., F.L.S. (of Eton College). Pp. xv., 322. 20 Plates. [With Preface by Dr. H. Woodward, Table of Contents, General Index, Explanations of the Plates, &c.] 1886, 4to. 25*s*.

The Genera and Species of Blastoidea, with a List of the Specimens in the British Museum (Natural History). By F. A. Bather, M.A., F.G.S., of the Geological Department. Pp. x., 70. 1 Woodcut. 1899, 8vo. 3*s*.

Catalogue of the Palæozoic Plants in the Department of Geology and Palæontology, British Museum (Natural History). By Robert Kidston, F.G.S. Pp. viii., 288. [With a list of works quoted, and an Index.] 1886, 8vo. 5*s*.

Catalogue of the Mesozoic Plants in the Department of Geology, British Museum (Natural History). By A. C. Seward, M.A., F.R.S., F.G.S., University Lecturer in Botany and Fellow of Emanuel College, Cambridge :—

Part I. The Wealden Flora. Part I. Thallophyta—Pteridophyta. Pp. xxxviii., 179. 17 Woodcuts and 11 Plates. [With Preface by Dr. Woodward, Alphabetical Index, Explanations of the Plates, &c.] 1894, 8vo. 10*s*.

Part II. The Wealden Flora. Part II. Gymnospermæ. Pp. viii., 259. 9 Woodcuts and 20 Plates. [With Alphabetical Index, Explanations of the Plates, &c.] 1895, 8vo. 15*s*.

Part III. The Jurassic Flora. Part I. The Yorkshire Coast. Pp. xii., 341. 53 Woodcuts and 21 Plates. [With Preface by Dr. Woodward, Alphabetical Index Explanations of the Plates, &c.] 1900, 8vo. 20*s*.

GUIDE-BOOKS.

(*To be obtained only at the Museum.*)

A General Guide to the British Museum (Natural History) Cromwell Road, London, S.W. With 31 Woodcuts, 2 Plans, 2 views of the building, and an illustrated cover. Pp. 97. 1901, 8vo. 3*d*. (*Out of print during revision.*)

Guide to the Galleries of Mammalia in the Department of Zoology of the British Museum (Natural History). 7th Edition. Pp. 126. 65 Woodcuts and 4 Plans. Index. 1902, 8vo. 6*d*.

Guide to the Galleries of Reptiles and Fishes in the Department of Zoology of the British Museum (Natural History). 4th Edition. Pp. iv., 119. 101 Woodcuts. Index. 1898. 8vo. 6*d.* (*Out of print.*)

Guide to the Shell and Starfish Galleries (Mollusca, Polyzoa, Brachiopoda, Tunicata, Echinoderma, and Worms), Department of Zoology, British Museum (Natural History). New Edition. Pp. iv., 130. 125 Woodcuts, Plan, and Indexes. 1901, 8vo. 6*d.*

Guide to the Coral Gallery (Protozoa, Porifera or Sponges, Hydrozoa, and Anthozoa) in the Department of Zoology, British Museum (Natural History). Pp. [iv.,] 73. 82 Illustrations, Plan, and Index, 1902, 8vo. 1*s.*

A Guide to the Fossil Mammals and Birds in the Department of Geology and Palæontology in the British Museum (Natural History). 7th Edition. [By Henry Woodward.] Pp. xii., 103. 116 Woodcuts. [With List of Illustrations, Table of Stratified Rocks, and Index.] 1896, 8vo. 6*d.* (*Out of print.*)

A Guide to the Fossil Reptiles and Fishes in the Department of Geology and Palæontology in the British Museum (Natural History). [By Henry Woodward.] Pp. xiv., 129 : 165 Woodcuts. [With List of Illustrations, Table of Stratified Rocks, and Index.] 1896, 8vo. 6*d.*

A Guide to the Fossil Invertebrates and Plants in the Department of Geology and Palæontology in the British Museum (Natural History). [By Henry Woodward.] Pp. xvi., 158. 182 Woodcuts. [With List of Illustrations, Table of Stratified Rocks, Introduction, and Index.] 1897, 8vo. 1*s.*

The same, in two parts :—

Part I. Mollusca to Bryozoa. Pp. xii., 64. 107 Woodcuts. [With List of Illustrations, Table of Stratified Rocks, and Introduction.] 1897, 8vo. 6*d.*

Part II. Insecta to Plants, &c. Pp. ix., 64*–158. Woodcuts 108–182. [With List of Illustrations and Index to the two parts.] 1897, 8vo. 6*d.*

Guide to Sowerby's Models of British Fungi in the Department of Botany, British Museum (Natural History). By Worthington G. Smith, F.L.S. Pp. 82. 93 Woodcuts. With Table of Diagnostic Characters, and Index. [Second Edition.] 1898, 8vo. 4*d.*

Guide to the British Mycetozoa exhibited in the Department of Botany, British Museum (Natural History). By Arthur Lister, F.L.S. Pp. 42. 44 Woodcuts. Index. 1903, 8vo. 3*d.*

A Guide to the Mineral Gallery of the British Museum (Natural History). [By L. Fletcher, M.A., F.R.S.] Pp. 32. Plan. 1903, 8vo. 1*d.*

The Student's Index to the Collection of Minerals, British Museum (Natural History). [By L. Fletcher, M.A., F.R.S.] Pp. 32. With a Plan of the Mineral Gallery. 1903, 8vo. 2*d.*

An Introduction to the Study of Minerals, with a Guide to the Mineral Gallery of the British Museum (Natural History). By L. Fletcher, M.A., F.R.S. Pp. 123. 41 Woodcuts. With Plan of the Mineral Gallery and Index. 1903, 8vo. 6*d.*

An Introduction to the Study of Rocks. By L. Fletcher, M.A., F.R.S. Pp. 118. [With plan of the Mineral Gallery, Table of Contents, and Index.] 1898, 8vo. 6*d.*

An Introduction to the Study of Meteorites, with a List of the Meteorites represented in the Collection. By L. Fletcher, M.A., F.R.S., &c. Pp. 95. [With a Plan of the Mineral Gallery, and an Index to the Meteorites represented in the Collection.] 1896, 8vo. 6*d.*

E. RAY LANKESTER,
Director.

British Museum
(Natural History),
Cromwell Road,
London, S.W.

October 1st, 1903.

Printed in Dunstable, United Kingdom

91221350R00147